SpringerBriefs in Optimization

Series Editors

Sergiy Butenko
Mirjam Dür
Panos M. Pardalos
János D. Pintér
Stephen M. Robinson
Tamás Terlaky
My T. Thai

SpringerBriefs in Optimization showcases algorithmic and theoretical techniques, case studies, and applications within the broad-based field of optimization. Manuscripts related to the ever-growing applications of optimization in applied mathematics, engineering, medicine, economics, and other applied sciences are encouraged.

More information about this series at http://www.springer.com/series/8918

Yaroslav D. Sergeyev · Dmitri E. Kvasov

Deterministic Global Optimization

An Introduction to the Diagonal Approach

Springer

Yaroslav D. Sergeyev
Dipartimento di Ingegneria Informatica,
   Modellistica, Elettronica e Sistemistica
Università della Calabria
Rende (CS)
Italy

and

Department of Software
   and Supercomputing Technologies
Lobachevsky State University
Nizhny Novgorod
Russia

Dmitri E. Kvasov
Dipartimento di Ingegneria Informatica,
   Modellistica, Elettronica e Sistemistica
Università della Calabria
Rende (CS)
Italy

and

Department of Software
   and Supercomputing Technologies
Lobachevsky State University
Nizhny Novgorod
Russia

ISSN 2190-8354   ISSN 2191-575X   (electronic)
SpringerBriefs in Optimization
ISBN 978-1-4939-7197-8   ISBN 978-1-4939-7199-2   (eBook)
DOI 10.1007/978-1-4939-7199-2

Library of Congress Control Number: 2017941484

Mathematics Subject Classification (2010): 90C26, 65K05, 93B30, 94A12

Printed on acid-free paper

This Springer imprint is published by Springer Nature
The registered company is Springer Science+Business Media LLC
The registered company address is: 233 Spring Street, New York, NY 10013, U.S.A.

Preface

A book should be luminous not voluminous.

Christian Nevell Bovee

This brief (non-voluminous) book is dedicated to deterministic global optimization dealing with optimization models characterized by the objective functions with several local optima (typically, their number is unknown and can be very high). Since the best set of parameters should be determined for these multiextremal models, traditional local optimization techniques and many heuristic approaches can be inappropriate and, therefore, global optimization methods should be applied. Moreover, in these problems the objective functions and constraints to be examined are often *black-box* and hard to evaluate. For example, their values can be obtained by executing some computationally expensive simulation, by performing a series of experiments, and so on. Such a kind of problems is frequently met in various fields of human activity (e.g., automatics and robotics, structural optimization, safety verification problems, engineering design, network and transportation problems, mechanical design, chemistry and molecular biology, economics and finance, data classification, etc.) and corresponds to computationally challenging global optimization problems, being actively studied around the world.

To obtain reliable estimates of the global solution based on a finite number of functions evaluations, some suppositions on the structure of the objective function and constraints (such as continuity, differentiability, convexity, and so on) should be indicated. These assumptions play a crucial role in the construction of any efficient global search algorithm, able to outperform simple uniform grid techniques in solving multiextremal problems. In fact, it is well known that if no particular assumptions are made on the objective function and constraints, any finite number of function evaluations does not guarantee getting close to the global minimum value of the objective function, since this function may have very high and narrow peaks.

One of the natural and powerful (from both the theoretical and the applied points of view) assumptions on the global optimization problem is that the objective function and constraints have bounded slopes. In other words, any limited change in the object parameters yields some limited changes in the characteristics of the objective performance. This assumption can be justified by the fact that in technical systems the energy of change is always bounded. One of the most popular mathematical formulations of this property is the Lipschitz continuity condition, which assumes that the absolute difference of any two function values is majorized by the difference of the corresponding function arguments (in the sense of a chosen norm), multiplied by a positive factor $L < \infty$. In this case, the function is said to be *Lipschitzian* and the corresponding factor L is said to be *Lipschitz constant*. The problem involving Lipschitz functions (the objective function and constraints) is said to be *Lipschitz global optimization problem*.

This brief book is dedicated to deterministic global optimization methods based on the Lipschitz condition. Multiextremal continuous problems with an unknown structure (black-box) with Lipschitz objective functions and functions having the first Lipschitz derivatives defined over hyperintervals are taken into consideration. Such problems arise very frequently in electrical and electronic engineering and other kinds of engineering applications. A brief survey of derivative-free methods and global optimization methods using the first derivatives is given for both one-dimensional and multidimensional cases. Algorithms using several techniques to balance local and global information during the search are described. A class of *diagonal algorithms* based on an efficient strategy that is applied for partitioning the search domain is described. Non-smooth and smooth minorants and acceleration techniques that can speed up several kinds of global optimization methods are introduced and investigated. Convergence conditions, examples of applications, numerical examples, and illustrations are provided. The book is essentially self-contained and is based mainly on results of the authors made in cooperation with a number of colleagues working in several countries.

The authors would like to thank the institutions they work at: University of Calabria, Italy; Lobachevsky State University of Nizhny Novgorod, Russia, and the Institute of High Performance Computing and Networking of the National Research Council of Italy. During the recent years the authors' research was supported by Italian and Russian Ministries of University, Education and Science and by the Italian National Institute of High Mathematics "F. Severi". Actually research activities of the authors are supported by the Russian Science Foundation, project num. 15-11-30022 "Global optimization, supercomputing computations, and applications".

The authors are very grateful to friends and colleagues for their inestimable help and useful and pleasant discussions: K. Barkalov, S. Butenko, V. Gergel, J. Gillard, S. Gorodetsky, V. Grishagin, D. Lera, M. Mukhametzhanov, P. Pardalos, R. Paulavičius, R. Strongin, A. Zhigljavsky, A. Žilinskas, and J. Žilinskas.

The continuous benevolent support of the Springer Editor R. Amzad is greatly appreciated.

The authors conclude this preface with cordial thanks to their families for their love and continuous support during the preparation of this book and not only.

Rende (CS), Italy/Nizhny Novgorod, Russia   Yaroslav D. Sergeyev
                                                             Dmitri E. Kvasov

Contents

Chapter 1
Lipschitz Global Optimization

A problem well stated is a problem half solved.

Charles F. Kettering

1.1 Problem Statement

In various fields of human activity (engineering design, economic models, biology studies, etc.) there arise different decision-making problems that can be often modeled by applying optimization concepts and tools. The decision-maker (engineer, physicist, chemist, economist, etc.) frequently wants to find the best combination of a set of parameters (geometrical sizes, electrical and strength characteristics, etc.) describing a particular optimization model which provides the optimum (minimum or maximum) of a suitable objective function. Often the satisfaction of a certain collection of feasibility constraints is required. The objective function usually expresses an overall system performance, such as profit, utility, risk, error, etc. In turn, the constraints originate from physical, technical, economic, or some other considerations (see, e.g., [10, 23, 103, 137, 211, 227, 242, 244, 290, 323, 329] and references given therein).

The sophistication of the mathematical models describing objects to be designed, which is a natural consequence of the growing complexity of these objects, significantly complicates the search for the best objective performance parameters. Many practical applications (see, e.g., [3, 10, 23, 71, 83, 90, 97, 101, 103, 130, 137, 148, 165, 209, 211, 214, 227, 228, 242, 244, 310, 311, 323, 330] and references given therein) are characterized by optimization problems with several local solutions (typically, their number is unknown and can be very high). These complicated problems are usually referred to as *multiextremal* problems (see, e.g., [129, 148, 150, 181, 231, 242, 290, 323, 348]). An example of a two-dimensional multiextremal objective function (from the class of functions considered in [132]) is given in Fig. 1.1; this type of objective functions is encountered, for instance, when the maximal working

© The Author(s) 2017
Y.D. Sergeyev and D.E. Kvasov, *Deterministic Global Optimization*,
SpringerBriefs in Optimization, DOI 10.1007/978-1-4939-7199-2_1

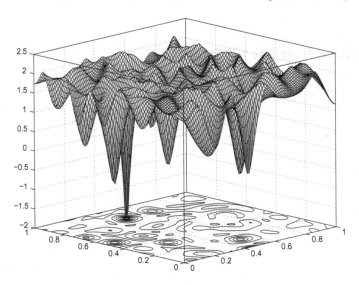

Fig. 1.1 A multiextremal two-dimensional objective function

stress over the thin elastic plate under some lumped transfer loads is estimated (see [323]).

Nonlinear local optimization methods and theory (described in a vast literature, see, e.g., [28, 49, 53, 68–70, 94, 143, 162, 201, 223, 260] and references given therein) not always are successful for solving these multiextremal problems. As indicated, for example, in [150, 323], their limitation is due to the intrinsic multiextremality of the problem formulation and not to the lack of smoothness or continuity—such properties are often present. It should be also mentioned in this context that in some problems a local solution may be just inappropriate. For example, if the objective function is an indicator of reliability and one is interested in assessing the worst case, then it is necessary to find the best (in this case, the worst) *global* solution (see, e.g., [290, 324]).

The theory, implementation, and application of models and strategies to solve multiextremal optimization problems are developed in the field called *global optimization*. Global optimization techniques are often significantly different from nonlinear local methods, and they can be (and are) much more expensive from the computational point of view. A great number of practical decision-making problems described by multiextremal objective functions and constraints and the rapid increase of computer power in the last decades determine a growing attention of the scientific community to global optimization. This attention results in developing a lot of approaches applied for solving global optimization problems and described in a vast literature (see, e.g., various monographs and surveys as [7, 9, 10, 23, 32, 66, 76, 78, 80, 82, 95, 99, 100, 102, 148–150, 156, 166, 167, 191, 197, 199, 206, 210, 212, 225, 228, 230, 242, 248, 252, 253, 258, 267, 269, 303, 307, 309, 315, 323, 327, 329, 332, 346, 347, 350, 353–355] and references given therein). It is important to notice here that global optimization methods can be completed by a local optimization phase aimed to refine a found estimate of the global solution (see, e.g., [67, 72, 150, 157, 242]). Global convergence, however, should be guaranteed by the global part of such 'combined' algorithms—otherwise, there exists a risk of missing the global solution. At the same time, it should be clear that a global approach that is too expensive (even having strong theoretical convergence properties and being straightforward to implement and apply) can place an excessively high demand on computational resources.

In practical applications, the objective function and constraints are often *black-box functions*. This means that there are 'black-boxes' associated with each of these functions that, with the values of the parameters given in input, return the corresponding values of the functions and nothing is known to the optimizer about the way these values are obtained. Such functions are met in many engineering applications in which observation of the produced function values can be made, but analytical expressions of the functions are not available. For example, the values of the objective function and constraints can be obtained by running some computationally expensive numerical model, by performing a set of experiments, etc. One may refer, for instance, to various decision-making problems in automatic control and robotics, structural optimization, safety verification problems, engineering design, network and transportation problems, mechanical design, chemistry and molecular biology, economics and finance, data classification, etc. (see, e.g., [16, 17, 27, 30, 33, 35, 41, 73, 84, 91, 103, 137, 138, 155, 160, 171, 196, 202, 211, 213, 242, 244, 256, 261, 290, 305, 323, 329, 334, 335, 337] and references given therein).

Due to the high computational costs involved, typically a small number of functions evaluations are available for a decision-maker (engineer, physicist, chemist, economist, etc.) when optimizing such costly functions. Thus, the main goal is to develop fast global optimization methods that produce acceptable solutions with a limited number of functions evaluations. However, to reach this goal, there are a lot of difficulties that are mainly related either to the lack of information about the objective function (and constraints, if any) or to the impossibility to adequately represent the available information about the functions.

For example, gradient-based algorithms (see, e.g., [102, 148, 228]) cannot be used in many applications because black-box functions are either non-differentiable or derivatives are not available and their finite-difference approximations are too expensive to obtain. Automatic differentiation (see, e.g., [54]), as well as interval-analysis-based approaches (see, e.g., [59, 156]), cannot be appropriately used in cases of black-box functions when their source codes are not available. A simple alternative could be the usage of the so-called direct (or derivative-free) search methods (see, e.g., [53, 61, 62, 157, 162, 195, 339]), frequently used now for solving engineering design problems (see, e.g., the DIRECT method [102, 154, 157], the response surface, or surrogate model methods [155, 256], etc.). However (see, e.g., [74, 179, 289]), these methods either are designed to find stationary points only or can require too high computational efforts for their work.

Therefore, solving the described global optimization problems is actually a real challenge both for theoretical and applied scientists. In this context, deterministic global optimization presents itself as a well developed mathematical theory having many important applications (see, e.g., [95, 102, 148, 242, 290, 323]). One of its main advantages is the possibility to obtain guaranteed estimates of global solutions and to demonstrate (under certain analytical conditions) rigorous global convergence properties. However, the currently available deterministic models can still require large number of functions evaluations to obtain adequately good solutions for these problems.

Stochastic approaches (see, e.g., [102, 148, 211, 228, 346, 348, 356]) can often deal with the stated problems in a simpler manner with respect to deterministic algorithms. They are also suitable for the problems where evaluations of the functions are corrupted by noise. However, there can be difficulties with these methods, as well (e.g., in studying their convergence properties if the number of iterations is finite). Several restarts are often required and this can lead to a high number of expensive functions evaluations, as well. Moreover, solutions found by some stochastic algorithms (especially, by heuristic methods like evolutionary algorithms, simulated annealing, etc.; see, e.g., [144, 206, 228, 251, 265, 343]) can often be only local solutions to the problems, located far from the global ones (see, e.g., [172–174, 298]). This can limit the usage of methods of this kind in practice. That is why we concentrate our attention on deterministic approaches.

To obtain estimates of the global solution related to a finite number of functions evaluations, some suppositions on the structure of the objective function and constraints (such as continuity, differentiability, convexity, and so on) should be indicated. These assumptions play a crucial role in the construction of any efficient global search algorithm able to outperform simple uniform grid techniques in solving multiextremal problems. In fact, as observed, e.g., in [75, 140, 258, 308, 340], if no particular assumptions are made on the objective function and constraints, any finite number of function evaluations does not guarantee getting close to the global minimum value, since this function may have a very high and narrow peaks.

One of the natural and powerful (from both the theoretical and the applied points of view) assumptions on the global optimization problem is that the objective function and constraints have bounded slopes. In other words, any limited change in the object parameters yields some limited changes in the characteristics of the objective performance. This assumption can be justified by the fact that in technical systems the energy of change is always limited (see the related discussion in [315, 323]). One of the most popular mathematical formulations of this property is the Lipschitz continuity condition, which assumes that the difference (in the sense of a chosen norm) of any two function values is majorized by the difference of the corresponding function arguments, multiplied by a positive factor $L < \infty$. In this case, the function is said to be *Lipschitzian* and the corresponding factor L is said to be *Lipschitz constant*. The problem involving Lipschitz functions (the objective function and constraints) is called *Lipschitz global optimization problem*.

As observed in [308], if the only information about the objective function $f(x)$ is that it belongs to the class of Lipschitz functions and the Lipschitz constant is unknown, there does not exist any deterministic or stochastic algorithm that, after a finite number of function evaluations, is able to provide an accurate ε-estimate of the global minimum f^*. Obviously, this result is very discouraging since in practice the information about the Lipschitz constant is usually difficult to obtain. Nevertheless, the necessity to solve practical problems remains. That is why in this case instead of the theoretical statement

(P1) "Construct an algorithm able to stop in a given time and to provide an ε-approximation of the global minimum f^*" the more practical statement

(P2) "Construct an algorithm able to stop in a given time and to return the lowest value of $f(x)$ obtained by the algorithm"

is used.

Under the statement (P2), either the computed solution (possibly improved by a local method) is accepted by the users (engineers, physicists, chemists, etc.) or the global search is repeated with changed parameters of the algorithm (see the related discussion in [302]). In the following, both the statements (P1) and (P2) will be used.

Formally, a general Lipschitz global optimization problem can be stated as follows:

$$f^* = f(x^*) = \min \ f(x), \quad x \in \Omega \subset \mathbb{R}^N, \tag{1.1}$$

where Ω is a bounded set defined by the following formulae:

$$\Omega = \{x \in D : g_i(x) \leq 0, \ 1 \leq i \leq m\}, \tag{1.2}$$

$$D = [a, b] = \{x \in \mathbb{R}^N : a(j) \leq x(j) \leq b(j), \ 1 \leq j \leq N\}, \quad a, b \in \mathbb{R}^N, \tag{1.3}$$

with N being the problem dimension. In (1.1)–(1.3), the objective function $f(x)$ and the constraints $g_i(x)$, $1 \leq i \leq m$, are multiextremal, not necessarily differentiable, black-box and hard to evaluate functions that satisfy the Lipschitz condition over the search hyperinterval D:

$$|f(x') - f(x'')| \leq L\|x' - x''\|, \quad x', x'' \in D, \tag{1.4}$$

$$|g_i(x') - g_i(x'')| \leq L_i\|x' - x''\|, \quad x', x'' \in D, \quad 1 \leq i \leq m, \tag{1.5}$$

where $\| \cdot \|$ denotes, usually, the Euclidean norm (see, e.g., for some other norms used), L and L_i, $1 \leq i \leq m$, are the (unknown) Lipschitz constants such that $0 < L < \infty$, $0 < L_i < \infty$, $1 \leq i \leq m$.

Note also that, due to the multiextremal character of the constraints, the admissible region Ω can consist of disjoint, non-convex subregions. More generally, not all the constraints (and, consequently, the objective function too) are defined over the whole region D: namely, a constraint $g_{i+1}(x)$ (or the objective function $g_{m+1} := f(x)$), $1 \leq i \leq m$, can be defined only over subregions where $g_j(x) \leq 0, 1 \leq j \leq i$. In this case, the problem (1.1)–(1.5) is said to be *with partially defined constraints*. If $m = 0$ in (1.2), the problem is said to be *box-constrained*.

This kind of problems is very frequent in practice. Let us refer only to the following examples: general (Lipschitz) nonlinear approximation; solution of nonlinear equations and inequalities; calibration of complex nonlinear system models; black-box systems optimization; optimization of complex hierarchical systems (related, for example, to facility location, mass-service systems); etc. (see, e.g., [102, 148, 149, 242, 244, 290, 323, 348] and references given therein). Numerous algorithms have been proposed (see, e.g., [22, 76, 82, 102, 109, 113, 140, 148, 150, 154, 176, 204, 207, 242, 246, 272, 276, 290, 303, 306, 313, 315, 323, 329, 336, 350] and references given therein) for solving the problem (1.1)–(1.5).

Sometimes, the problem (1.1), (1.3), (1.4) with a differentiable objective function having the Lipschitz (with an unknown Lipschitz constant) first derivative $f'(x)$ (which could be itself a costly black-box function) is included in the same class of Lipschitz global optimization problems (theory and methods for solving this differentiable problem can be found, e.g., in [13, 15, 34, 39, 45, 86, 100, 116, 117, 177, 242, 275, 276, 278, 290, 303, 323]).

It should be stressed that not only multidimensional global optimization problems but also one-dimensional ones (in contrast to one-dimensional local optimization problems that have been very well studied in the past) continue to attract attention of many researchers. This happens at least for two reasons. First, there exist a large number of applications where it is necessary to solve such problems. Electrical engineering and electronics are among the fields where one-dimensional global optimization methods can be used successfully (see, e.g., [47, 48, 64, 65, 217, 282, 290, 323, 331]). Second, there exist numerous approaches (see, e.g., [100, 148, 150, 208, 242, 286, 290, 315, 323, 329, 346]) enabling to generalize methods proposed for solving univariate problems to the multidimensional case.

It is also worthy to notice that it is not easy to find suitable ways for managing multiextremal constraints (1.2). For example, the traditional penalty approach (see, e.g., [28, 92, 148, 150, 223, 343] and references given therein) requires that $f(x)$ and $g_i(x)$, $1 \le i \le m$, are defined over the whole search domain D. At first glance it seems that at the regions where a constraint is violated the objective function can be simply filled in with either a big number or the function value at the nearest feasible point (see, e.g., [148, 150, 153]). Unfortunately, in the context of Lipschitz algorithms, incorporating such ideas can lead to infinitely high Lipschitz constants, causing degeneration of the methods and non-applicability of the penalty approach (see the related discussion in [102, 242, 290, 323, 348]).

A promising approach called the *index scheme* has been proposed in [316, 321] (see also [18, 136, 294, 300, 323]) to reduce the general constrained problem (1.1)–(1.5)) to a box-constrained discontinuous problem. An important advantage of the index scheme is that it does not introduce additional variables and/or parameters by opposition to widely used traditional approaches. It has been shown in [283] that the index scheme can be also successfully used in a combination with the branch-and-bound approach (see, e.g., [150, 242, 277, 290]) if the Lipschitz constants from (1.4)–(1.5) are known a priori. An extension of the index approach, originally proposed for information stochastic Bayesian algorithms, to the framework of geometric algorithms constructing non-differentiable discontinuous minorants for the reduced problem, has been also examined (see, e.g., [281, 287, 294, 302]).

In this brief book, in order to expose the principal ideas of some known approaches to solving the stated problem, the box-constrained Lipschitz global optimization problem (1.1), (1.3), (1.4) will be considered. In particular, approaches under consideration will be distinguished by the mode in which information about the Lipschitz constants is obtained and by the strategy of exploration of the search hyperinterval D from (1.3).

1.2 Lipschitz Condition and Its Geometric Interpretation

The Lipschitz continuity assumption, being quite realistic for many practical problems (see, e.g., [82, 102, 148, 150, 180, 242, 244, 290, 291, 323, 348] and references given therein), is also an efficient tool for constructing global optimum estimates with a finite number of function evaluations. Moreover, it gives the possibility to construct global optimization algorithms and to prove their convergence and stability.

In fact, the Lipschitz condition allows us to give its simple geometric interpretation (see Fig. 1.2). Once a one-dimensional Lipschitz (with the Lipschitz constant L) function $f(x)$ has been evaluated at two points x' and x'', due to (1.4) the following four inequalities take place over the interval $[x', x'']$:

$$f(x) \leq f(x') + L(x - x'), \quad f(x) \geq f(x') - L(x - x'), \quad x \geq x',$$

$$f(x) \leq f(x'') - L(x - x''), \quad f(x) \geq f(x'') + L(x - x''), \quad x \leq x''.$$

Thus, the function graph over the interval $[x', x'']$ must lie inside the area limited by four lines which pass through the points $(x', f(x'))$ and $(x'', f(x''))$ with the slopes $\pm L$ (see the light gray parallelogram in Fig. 1.2).

If the function $f(x)$ is to be minimized over the interval $[x', x'']$, its lower bound over this interval can be easily calculated. In fact, due to the Lipschitz condition (1.4), the following inequality is satisfied:

$$f(x) \geq F(x),$$

where $F(x)$ is the following piecewise linear function over $[x', x'']$:

$$F(x) = \max\{f(x') - L(x - x'), f(x'') + L(x - x'')\}, \quad x \in [x', x''].$$

The minimal value of $F(x)$ over $[x', x'']$ is therefore the lower bound of $f(x)$ over this interval. It is calculated as

$$R = R_{[x',x'']} = F(\hat{x}) = \frac{f(x') + f(x'')}{2} - L\frac{x'' - x'}{2} \tag{1.6}$$

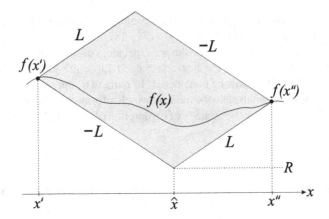

Fig. 1.2 Graphic interpretation of the Lipschitz condition

and is obtained (see Fig. 1.2) at the point

$$\hat{x} = \frac{x' + x''}{2} - \frac{f(x'') - f(x')}{2L}. \tag{1.7}$$

The described geometric interpretation lies in the basis of the Piyavskij–Shubert method (see [246, 306])—one of the first and well studied algorithms for the one-dimensional Lipschitz global optimization, discussed in a vast literature (see, e.g., the survey [140] and references given therein). Various modifications of this algorithm have been also proposed (see, e.g., [22, 82, 140, 150, 266, 304, 328]). As observed, e.g., in [63, 152, 325], this algorithm is 'optimal in one step', i.e., the choice of the current evaluation point as in (1.7) ensures the maximal increase of the lower bound of the global minimum value of $f(x)$ with respect to a number of reasonable criteria (see, e.g., [150, 323, 326, 348] for the related discussions on optimality principles in Lipschitz global optimization).

The Piyavskij–Shubert method belongs to the class of *geometric algorithms*, i.e., algorithms that use in their work auxiliary functions to estimate behavior of $f(x)$ over the search region, as just described above. This idea has proved to be very fruitful and many algorithms (both one-dimensional and multidimensional ones; see, e.g., [13, 22, 45, 109, 140, 147, 159, 207, 224, 328, 341, 344] and references given therein) based on constructing, updating, and improving auxiliary piecewise minorant functions built using an overestimate of the Lipschitz constant L have been proposed. Close geometric ideas have been used for other classes of optimization problems, as well. For example, various aspects of the so-called αBB-algorithm for solving constrained non-convex optimization problems with twice-differentiable functions have been examined, e.g., in [2, 6, 202] (see also [45, 96, 97]). The αBB-algorithm combines a convex lower bounding procedure (based on a parameter α that can be interpreted in a way similar to the Lipschitz constant) within the branch-and-bound framework whereas nonconvex terms are underestimated through quadratic functions derived from a second-order information.

As shown, e.g., in [216, 290, 323], there exists a strong relationship between the geometric approach and another powerful technique for solving the stated problem— the so-called *information-statistical approach*. This approach takes its origins in the works [220, 313] (see also [315, 323]) and, together with Piyavskij–Shubert method, it has consolidated foundations of the Lipschitz global optimization. The main idea of this approach is to apply the theory of random functions to building a mathematical representation of an available (certain or uncertain) a priori information on the objective function behavior. A systematic approach to the description of some uncertain information on this behavior is to accept that the unknown black-box function to be minimized is a sample of some known random function. Generally, to provide an efficient analytical technique for deriving some estimates of global optimum with a finite number of *trials* (function evaluations), i.e., for obtaining by Bayesian reasoning some conditional (with respect to the trials performed) estimations, the random function should have some special structure. Then, it is possible to deduce the decision rules for performing new trials as optimal decision functions (see, e.g., [29, 38, 114, 155, 186, 210, 212, 315, 323, 329, 345, 348–350]).

One of the main questions to be considered in solving the Lipschitz global optimization problem is, How can the Lipschitz constant L of the objective function $f(x)$ from (1.1) be obtained and used? There are several approaches to specify the Lipschitz constant. For example, an overestimate $\hat{L} \geq L$ can be given a priori (see, e.g., [22, 81, 85, 140, 148, 150, 204, 207, 246, 306, 341]). This case is very important from the theoretical viewpoint but $\hat{L}$ is usually not easily obtainable in practice. More flexible approaches are based on an adaptive estimation of L in the course of the search. In such a way, algorithms can use either a *global* estimate of the Lipschitz constant (see, e.g., [150, 176, 219, 242, 313, 315, 323]) valid for the whole region D from (1.3), or *local* estimates L_i valid only for some subregions $D_i \subseteq D$ (see, e.g., [175, 216, 272, 273, 290, 323]). Estimating local Lipschitz constants during the work of a global optimization algorithm allows one to significantly accelerate the global search (the importance of the estimation of local Lipschitz constants has been highlighted also in other works, see, e.g. [13, 205, 242, 320]). Naturally, balancing between local and global information must be performed in an appropriate way (see, e.g., [272, 290, 323]) to avoid the missing of the global solution. Finally, estimates of L can be chosen during the search from a certain set of admissible values from zero to infinity (see, e.g., [93, 107, 141, 153, 154, 289, 290]). It should be stressed that either the Lipschitz constant L is known and an algorithm is constructed correspondingly, or L is not known but there exist sufficiently large values of parameters of the considered algorithm ensuring its convergence (see, e.g., (see, e.g., [290, 303, 323]).

Note that the Lipschitz constant has also a significant influence on the convergence speed of numerical algorithms and, therefore, the problem of its adequate specifying is of the great importance for the Lipschitz global optimization (see, e.g., [119, 140, 149, 154, 205, 242, 272, 273, 290, 323, 342, 348]). An underestimate of the Lipschitz constant L can lead to the loss of the global solution. In contrast, accepting a very high value of an overestimate of L for a concrete objective function means

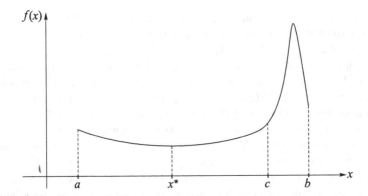

Fig. 1.3 A function with a high global Lipschitz constant

(due to the Lipschitz condition (1.4)) assuming that the function has a complicated structure with sharp peaks and narrow attraction regions of minimizers within the whole admissible region. Thus, a too high overestimate of L (if it does not correspond to the real behavior of the objective function) leads to a slow convergence of the algorithm to the global minimizer.

In Fig. 1.3 (see [323]), an example of a one-dimensional function with a high value of the global Lipschitz constant L over the feasible interval $[a, b]$ is shown. This value of the Lipschitz constant is obtained over the subinterval $[c, b]$. The global minimizer x^* belongs to the wide subinterval $[a, c]$, where the function has a low local Lipschitz constant. In this situation, a global optimization method using the high global constant L (or its overestimate) will not take into account any local information (represented by the local Lipschitz constants) about the behavior of the objective function over the subinterval $[a, c]$. Therefore, such a method will work very slowly at the interval $[a, c]$, in spite of the simplicity of the objective function and the low local Lipschitz constant over this interval. Thus, behavior of the method in a wide neighborhood of x^* depends on the local constant over the interval $[c, b]$ (which is also the global constant) being not only narrow but also lying far away from the global minimizer x^* and corresponding to the global maximum that is out of our interest. As has been demonstrated in [188, 272, 273, 323], estimating local Lipschitz constants during the work of a global optimization algorithm allows one to significantly accelerate the global search (see also, e.g. [13, 205, 242]). Naturally, a smart balancing between local and global information must be performed in an appropriate way to avoid the missing of the global solution.

1.3 Multidimensional Approaches

Since the uniform grid techniques require a substantial number of evaluation points, they can be successfully applied only if the dimensionality N from (1.3) is small or the required accuracy is not high. As observed in [148, 242, 317, 323], an adaptive selection of evaluation points $x^1, x^2, \ldots, x^k$ can outperform the grid technique. Such a selection aims to ensure a higher density of the evaluation points near the global minimizer x^* with respect to the densities in the rest of the search domain D.

One of these adaptive approaches is a direct extension of the one-dimensional geometric algorithms to the multidimensional case. At each iteration $k \geq 1$ of such an algorithm, a function

$$F_k(x) = \max_{1 \leq i \leq k} \{f(x^i) - \hat{L}\|x - x^i\|\}, \quad \hat{L} > L, \quad x \in D, \tag{1.8}$$

is constructed where $F_k(x)$ is a lower bounding function (minorant) for $f(x)$ (due to the Lipschitz condition (1.4) and the fact that $\hat{L}$ is an overestimate of the constant L), i.e.,

$$f(x) \geq F_k(x), \quad \forall k \geq 1, \quad x \in D.$$

A global minimizer of the function (1.8) is then determined and chosen as a new evaluation point for $f(x)$, i.e.,

$$x^{k+1} = \arg\min_{x \in D} F_k(x). \tag{1.9}$$

After evaluation of $f(x^{k+1})$, the minorant function (1.8) is updated and the process is repeated until a stopping criterion is satisfied. For example, the algorithm can be stopped when the difference between the current estimate of the minimal function value and the global minimum of $F_k(x)$, which is a lower bound on the global minimum of $f(x)$, becomes smaller than some prescribed value.

Geometrically, the graph of $F_k(x)$ in $\mathbb{R}^{N+1}$ is defined by a set of k intersecting N-dimensional cones bounded by the hyperinterval $D \subset \mathbb{R}^N$, with vertices $(x^i, f(x^i))$, $1 \leq i \leq k$, and the slope $\hat{L}$. Axes of symmetry of these cones are parallel to the $(N+1)$-th unit vector and their sections orthogonal to the axes are spheres (see Fig. 1.4). In the case of one dimension ($N = 1$), there are no serious problems in calculating x^{k+1} from (1.9). In fact, this is reduced to selecting the minimal value among k values, each of which is easy to compute due to the explicit formula (1.6) (see Fig. 1.2). If $N > 1$, then the relations between the location of the next evaluation point and the results of the already performed evaluations become significantly more complicated. Therefore, finding the point x^{k+1} can become the most time-consuming operation of the multidimensional algorithm, and its complexity increases with the growth of the problem dimension.

In [140, 246], it is observed that the global minimizer x^{k+1} of the minorant function $F_k(x)$ belongs to a finite set of points, i.e., local minima of $F_k(x)$, which can be computed by solving systems of quadratic equations. In [207], a deep analysis of the algorithm constructing lower bounding functions $F_k(x)$ from (1.8) is presented and a more efficient way to find their local minima is proposed, which requires solving systems of N linear and one quadratic equations. Algebraic methods aiming to avoid computing solutions of some systems which do not correspond to global minima of

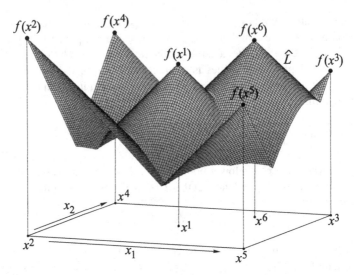

Fig. 1.4 An example of a two-dimensional lower bounding function for an objective function $f(x)$, $x = (x_1, x_2) \in D \subset \mathbb{R}^2$, evaluated at the points $x^1, \ldots, x^6$

$F_k(x)$ are considered in [312]. In [140], it is indicated that the number of systems to be examined can be further reduced.

In [341], cones with the spherical base (1.8) are replaced by cones with a simplex base. At each iteration, there exists a collection of such inner approximating simplices in $\mathbb{R}^{N+1}$ that bound parts of the graph of the objective function $f(x)$ from below, always including the part containing the global minimum point (x^*, f^*) from (1.1). Various aspects of generalizations of this technique are discussed in [13, 344].

In [203], minorant functions similar to (1.8) are considered. The solution of (1.9) still remains difficult, the problem of the growth of complexity of (1.9) is analyzed, and a strategy for dropping elements from the sequence of evaluation points while constructing a minorant function is proposed.

Unfortunately, the difficulty of determining a new evaluation point (1.9) in high dimensions (after executing even relatively small number of function evaluations) has not been yet overcome. It should be noted here that, as observed in [323], this difficulty is usual for many approaches aiming to decrease the total number of evaluations by constructing adaptive auxiliary functions approximating $f(x)$ over the search domain D. Methods of this kind introduce some selection rule of the type (1.9) and an auxiliary function similar to $F_k(x)$ is used in such a rule. This function often happens to be essentially multiextremal (also in cases where its construction is connected with assumptions quite different from the Lipschitz condition (1.4)). Moreover, in this case, it is difficult to use adaptive estimates of the global or local Lipschitz constant(s). This fact can pose a further limitation for practical methods directly extending the one-dimensional Piyavskij–Shubert algorithm to the multidimensional case.

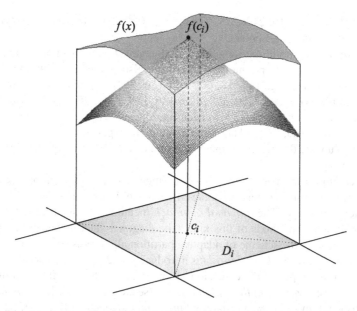

Fig. 1.5 The objective function is evaluated only at the central point c_i of a hyperinterval D_i and the construction of the bounding cone is performed over D_i independently of the other hyperintervals

In order to simplify the task of calculating lower bounds of $f(x)$ and determining new evaluation points, many multidimensional methods subdivide the search domain $D = [a, b]$ from (1.3) into a set of subdomains D_i. In this way, at each iteration k a partition $\{D^k\}$ of D is obtained:

$$D = \bigcup_{i=1}^{M} D_i, \quad D_i \cap D_j = \delta(D_i) \cap \delta(D_j), \quad i \neq j, \qquad (1.10)$$

where $M = M(k)$ is the number of subdomains D_i present during the kth iteration and $\delta(D_i)$ denotes the boundary of D_i. An approximation of $f(x)$ over each subregion D_i from $\{D^k\}$ is based on the results obtained by evaluating $f(x)$ at trial points $x \in D$.

If a partition of D into hyperintervals D_i is used in (1.10) and the objective function is evaluated, for example, at the central point of every hyperinterval D_i (the so-called *center-sampling partition strategies*—see, e.g., [107, 109, 153, 154, 204, 205, 280]), each cone from (1.8) can be considered over the corresponding hyperinterval D_i, independently of the other cones (see Fig. 1.5). This allows one to avoid the necessity of establishing the intersections of the cones and to simplify the lower bound estimation of $f(x)$ when a new evaluation point is determined by a rule like (1.9). The more accurate estimation of the objective function behavior can be achieved when two evaluation points over a hyperinterval are used for constructing an auxiliary function for $f(x)$, for example, at the vertices corresponding to the main diagonal of hyperintervals (the so-called *diagonal partition strategies*—see, e.g.,

[117, 150, 170, 176, 216, 242, 278, 279, 289, 290, 292]). Methods of this type are the main topic of this brief book.

In exploring the multidimensional search domain D, various multisection techniques for partitions of hyperintervals have been also studied in the framework of interval analysis (see, e.g., [46, 58, 60, 139, 156, 254]). More complex partitions based on simplices (see, e.g., [50, 127, 129, 146, 150, 234, 235, 237]) and auxiliary functions of various nature have also been introduced (see, e.g., [5, 8, 13, 25, 26, 34, 36, 105, 161, 163, 164, 341]). Several attempts to generalize various partition schemes in a unique theoretical framework have been made (see, e.g., [88, 150, 195, 242, 277, 327]).

Another fruitful approach to solving multidimensional Lipschitz global optimization problems consists of extending some efficient one-dimensional algorithms to the multidimensional case. The *diagonal approach* mentioned above (see [242, 290]) is the first group of methods of this kind. The diagonal algorithms can be viewed as a special case of the scheme of adaptive partition algorithms (or a more general scheme of the *divide-the-best algorithms* introduced in [277, 290]). There also exist at least two other types of the extension of univariate methods to the multidimensional case: *nested global optimization scheme* (see, e.g., [43, 246]) and *reduction of the dimension by using Peano space-filling curves* (see, e.g., [303, 315, 323]).

In the nested optimization scheme (see [43, 246] and also [115, 118, 129, 134, 286, 315, 323]), the minimization over D of the Lipschitz function $f(x)$ from (1.1) is made using the following nested form

$$\min_{x \in D} f(x) = \min_{a(1) \le x(1) \le b(1)} \cdots \min_{a(N) \le x(N) \le b(N)} f(x(1), \ldots, x(N)).$$

This allows one to solve the multidimensional problem (1.1), (1.3), (1.4) by solving a set of nested univariate problems and, thus, to apply methods proposed for solving the one-dimensional optimization problems.

The second dimensionality reduction approach (see, e.g., [4, 37, 303, 315, 318, 323, 324]) is based on the application of a single-valued Peano-type curve $x(\tau)$ continuously mapping the unit interval [0, 1], $\tau \in [0, 1]$, onto the hypercube D. These curves, introduced by Peano and further by Hilbert, pass through any point of D, i.e., 'fill in' the hypercube D, and this gave rise to the term *space-filling curves* (see [263, 319]) or just *Peano curves*. Particular schemes for the construction of various approximations to Peano curves, their theoretical justification, and the standard routines implementing these algorithms are given, e.g., in [303, 315, 317, 322, 323].

If $x(\tau)$ is Peano curve, then from the continuity (due to (1.4)) of $f(x)$ it follows that

$$\min_{x \in D} f(x) = \min_{\tau \in [0,1]} f(x(\tau)) \tag{1.11}$$

and this reduces the original multidimensional problem (1.1), (1.3), (1.4) to one dimension (see Fig. 1.6). Moreover, it can be proved (see, e.g., [323]) that if the mul-

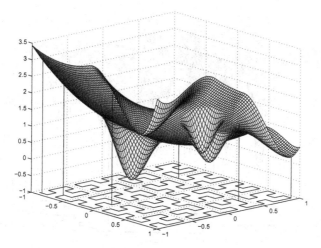

Fig. 1.6 Usage of an approximation of the Peano curve for global optimization: A two-dimensional function from the class [111] has been evaluated at six points belonging to the curve

tidimensional function $f(x)$ from (1.1) is Lipschitzian with a constant L, then the reduced one-dimensional function $f(x(\tau))$, $\tau \in [0, 1]$, satisfies the uniform Hölder condition over $[0, 1]$ with the exponent N^{-1} and the coefficient $2L\sqrt{N+3}$, i.e.:

$$|f(x(\tau')) - f(x(\tau''))| \leq 2L\sqrt{N+3}\,(|\tau' - \tau''|)^{1/N}, \quad \tau', \tau'' \in [0, 1]. \quad (1.12)$$

Thus, the Lipschitz assumption does not hold for the function $f(x(\tau))$ at the right-hand side of (1.11), but $f(x(\tau))$ satisfies the Hölder condition (1.12). As shown, e.g., in [131, 185, 187, 189, 303, 315, 323, 333], several Lipschitz one-dimensional algorithms can be successfully generalized to the case of minimization of Hölder functions, too. In the monographs [303, 323], many powerful sequential and parallel global optimization methods based on the usage of Peano curves are proposed and analyzed in depth.

Let us now return to the already mentioned diagonal approach (see, e.g., [241, 242, 279, 290] to solving Lipschitz global optimization problems. It provides that the initial hyperinterval D from (1.3) is partitioned into a set of smaller hyperintervals D_i, the objective function is evaluated only at two vertices corresponding to the main diagonal of hyperintervals D_i of the current partition of D, and the results of these evaluations are used to select a hyperinterval for the further subdivision. The diagonal approach has a number of attractive theoretical properties and has proved to be efficient in solving a number of applied problems (see, e.g., [120, 171, 178, 181, 182, 242, 289, 290] and references given therein).

Thus, an extension of one-dimensional global optimization algorithms to the multidimensional case can be performed naturally by means of the diagonal scheme (see, e.g., [170, 175, 176, 241, 242, 279, 289, 290]) that works as follows. In order to decrease the computational efforts needed to describe the objective function $f(x)$

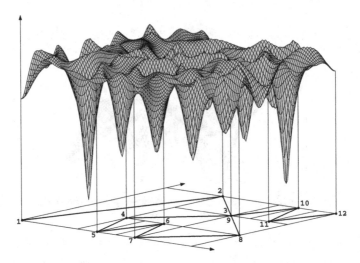

Fig. 1.7 Usage of adaptive diagonal curves (evaluation points from 1 to 12), generated by the efficient diagonal partition strategy, for minimizing a multiextremal two-dimensional objective function from the class [132]

at every small hyperinterval $D_i \subset D$, $f(x)$ is evaluated only at the vertices a_i and b_i corresponding to the main diagonal $[a_i, b_i]$ of D_i. At every iteration the 'merit' of each hyperinterval so far generated is estimated. The higher 'merit' of D_i corresponds to the higher possibility that the global minimizer x^* of $f(x)$ belongs to D_i. The 'merit' is measured by a real-valued function R_i called *characteristic* of the hyperinterval D_i (see [129, 135, 290, 323]). In order to calculate the characteristic R_i of a multidimensional hyperinterval D_i, some one-dimensional characteristics can be used as prototypes. After an appropriate transformation they can be applied to the one-dimensional segment being the main diagonal $[a_i, b_i]$ of D_i. A hyperinterval having the 'best' characteristic (e.g., the biggest one, as in the information approach, see, e.g., [315, 319, 323], or the smallest one, as in the geometric approach, see, e.g., [273, 290, 323]) is partitioned by means of a partition operator (*diagonal partition strategy*), and new evaluations are performed at two vertices corresponding to the main diagonal of each generated hyperinterval. Each concrete choice of the characteristic R_i and the partition strategy determines a particular diagonal method (see, e.g., [107, 150, 154, 176, 216, 242, 279, 289, 290] and references given therein).

Different exploration techniques based on various diagonal adaptive partition strategies have been analyzed in [290] (see also [242]). It has been demonstrated in [279, 290, 293] that partition strategies traditionally used in the framework of the diagonal approach may not fulfil requirements of computational efficiency because of the execution of many redundant evaluations of the objective function. Such a redundancy slows down significantly the global search in the case of costly functions.

A new diagonal adaptive partition strategy, originally proposed in [279] (see also [280, 290]), that allows one to avoid this computational redundancy has been thoroughly described in [290]. It can be also viewed as a procedure generating a series of curves similar to Peano space-filling curves—the so-called *adaptive diagonal curves*. It should be mentioned that the depth of partition of the adaptive diagonal curves is different within different subdomains of the search region. A particular diagonal algorithm developed by using the new partition strategy constructs its own series of curves, taking into account properties of the problem to be solved. The curves become more dense in the vicinity of the global minimizers of the objective function if the selection of hyperintervals for partitioning is realized appropriately (see Fig. 1.7). Thus, these curves can be used for reducing the problem dimensionality similarly to Peano curves but in a more efficient way.

In [290], it is shown both theoretically and numerically that the new strategy considerably speeds up the search and also leads to saving computer memory. It is particularly important that its advantages become more pronounced when the problem dimension N increases. Thus, fast one-dimensional methods can be efficiently extended to the multidimensional case by using this scheme. Moreover, the proposed partition strategy can be successfully parallelized by techniques from [135, 320, 323] spswhere theoretical results regarding non-redundant parallelism have been obtained, as wellsps. Clearly, such a non-redundant parallelization gives the possibility to obtain an additional speed up.

In conclusion to this introductory chapter it should be emphasized that a particular attention should be also paid to the problem of testing global optimization algorithms. The lack of such information as the number of local optima, their locations, attraction regions, local and global values, describing global optimization tests taken from real-life applications creates additional difficulties in verifying validity of the algorithms. Some benchmark classes and related discussions on testing global optimization algorithms can be found, e.g., in [1, 19, 24, 76, 77, 89, 98, 103, 108, 111, 132, 148, 184, 198, 222, 226, 259, 264, 268, 290, 297, 323, 348].

Chapter 2
One-Dimensional Algorithms and Their Acceleration

It is common sense to take a method and try it. If it fails, admit it frankly and try another. But above all, try something.

Franklin D. Roosevelt

2.1 One-Dimensional Lipschitz Global Optimization

In order to give an insight to the class of geometric Lipschitz global optimization (LGO) methods, one-dimensional problems will be considered in this chapter. In global optimization, these problems play a very important role both in the theory and practice and, therefore, they were intensively studied in the last decades (see, e.g., [51, 82, 102, 148, 242, 290, 315, 323, 326, 348]). In fact, on the one hand, theoretical analysis of one-dimensional problems is quite useful since mathematical approaches developed to solve them very often can be generalized to the multidimensional case by numerous schemes (see, e.g., [58, 148, 150, 154, 176, 187, 207, 227, 234, 242, 278, 290, 323, 348]). On the other hand, there exist a large number of real-life applications where it is necessary to solve these problems (see, e.g., [227, 242, 244, 257, 290, 323, 348]).

Let us consider, for example, the following common problem in electronic measurements and electrical engineering. There exists a device whose behavior depends on a characteristic $f(x)$, $x \in [a, b]$, where the function $f(x)$ may be, for instance, an electrical signal obtained by a complex computer aided simulation over a time interval $[a, b]$ (see the function graph shown by thick line in Fig. 2.1). The function $f(x)$ is often multiextremal and Lipschitzian (it can be also differentiable with the Lipschitz first derivative). The device works correctly while $f(x) > 0$. Of course, at the initial moment $x = a$ we have $f(a) > 0$. It is necessary to describe the performance of the device over the time interval $[a, b]$ either by determining the point x^* such that

© The Author(s) 2017

Y.D. Sergeyev and D.E. Kvasov, *Deterministic Global Optimization*, SpringerBriefs in Optimization, DOI 10.1007/978-1-4939-7199-2_2

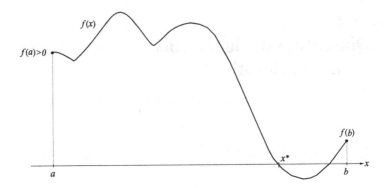

Fig. 2.1 The problem arising in electrical engineering of finding the minimal root of an equation $f(x) = 0$ with multiextremal non-differentiable *left* part

$$f(x^*) = 0, \quad f(x) > 0, \quad x \in [a, x^*), \quad x^* \in (a, b], \tag{2.1}$$

or by demonstrating that x^* satisfying (2.1) does not exist in $[a, b]$. In the latter case the device works correctly for the whole time period; thus, any information about the global minimum of $f(x)$ could be useful in practice to measure the device reliability.

This problem is equivalent to the problem of finding the minimal root (the first root from the left) of the equation $f(x) = 0$, $x \in [a, b]$, in the presence of certain initial conditions and can be reformulated as a global optimization problem. There is a simple approach to solve this problem based on a grid technique. It produces a dense mesh starting from the left margin of the interval and going on by a small step till the signal becomes less than zero. For an acquired signal, the determination of the first zero crossing point by this technique is rather slow especially if the search accuracy is high. Since the objective function $f(x)$ is multiextremal (see Fig. 2.1) the problem is even more difficult because many roots can exist in $[a, b]$ and, therefore, classical local root finding techniques can be inappropriate and application of global methods become desirable.

A box-constrained Lipschitz global optimization problem (1.1), (1.3), (1.4) can be stated in its one-dimensional version as follows. Given a small positive constant ε, it is required to find an ε-approximation of the global minimum point (*global minimizer*) x^* of a multiextremal, black-box (and, often, hard to evaluate) objective function $f(x)$ over a closed interval $[a, b]$:

$$f^* = f(x^*) = \min f(x), \quad x \in [a, b]. \tag{2.2}$$

It can be supposed either that the objective function $f(x)$ is not necessarily differentiable and satisfies the Lipschitz condition with an (unknown) Lipschitz constant L, $0 < L < \infty$,

$$|f(x') - f(x'')| \leq L|x' - x''|, \quad x', x'' \in [a, b], \tag{2.3}$$

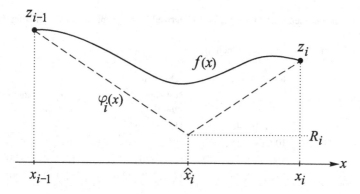

Fig. 2.2 Geometric interpretation of the Lipschitz condition

or that $f(x)$ is differentiable with the first derivative $f'(x)$ being itself multiextremal black-box Lipschitz function with an (unknown) Lipschitz constant $K, 0 < K < \infty$,

$$|f'(x') - f'(x'')| \leq K|x' - x''|, \quad x', x'' \in [a, b]. \tag{2.4}$$

As it was mentioned in Sect. 1.1, both the problems (2.2), (2.3) and (2.2), (2.4) are very frequent in practice (see, e.g., [10, 51, 102, 103, 137, 148, 227, 229, 242, 244, 255, 323, 348]).

The subsequent sections of this chapter will be dedicated to the geometric methods based on the geometric interpretation of the Lipschitz condition (2.3) or (2.4) for solving both the problems. Let us consider first the problem (2.2)–(2.3) and suppose that the objective function $f(x)$ from (2.2)–(2.3) has been evaluated at two points x_{i-1} and x_i of the search interval $[a, b]$ with the corresponding function values $z_{i-1} = f(x_{i-1})$ and $z_i = f(x_i)$ (see the function graph shown by thick line in Fig. 2.2). Then, the following inequality is satisfied over $[x_{i-1}, x_i]$:

$$f(x) \geq \phi_i(x),$$

where $\phi_i(x)$ is a piecewise linear function called *minorant* or *lower bounding function* (its graph is drawn by dashed line in Fig. 2.2),

$$\phi_i(x) = \max\{z_{i-1} - L(x - x_{i-1}), \ z_i + L(x - x_i)\}, \quad x \in [x_{i-1}, x_i]. \tag{2.5}$$

The minimal value of $\phi_i(x)$ over $[x_{i-1}, x_i]$ is therefore a lower bound of $f(x)$ over this interval. It is called *characteristic* (see [133, 135, 323]) of the interval $[x_{i-1}, x_i]$ and is calculated as follows:

$$R_i = R_{[x_{i-1}, x_i]} = \phi_i(\hat{x}_i) = \frac{z_{i-1} + z_i}{2} - L\frac{x_i - x_{i-1}}{2}, \tag{2.6}$$

where the point $\hat{x}_i$ (see Fig. 2.2) is

Table 2.1 Some geometric LGO methods characterized by different ways of estimating the Lipschitz constant

Lipschitz constant estimate	Objective function		
	Non-differentiable	Differentiable	
		Non-smooth minorants	Smooth minorants
A priori	[81, 140, 246, 306]	[13, 34, 116]	[200, 271, 276]
Multiple	[107, 141, 154, 289]	[177]	—
Global	[150, 242, 315, 323]	[116, 270, 276]	[271, 276, 323]
Local	[272, 273, 290, 323]	[270, 275, 323]	[271, 276, 290]

$$\hat{x}_i = \frac{x_{i-1} + x_i}{2} - \frac{z_i - z_{i-1}}{2L}. \tag{2.7}$$

In the next two sections, several geometric methods for solving both the problems (2.2), (2.3) and (2.2), (2.4) will be considered by classifying them on the way of obtaining the Lipschitz information (a priori given, multiple, global, and local estimates of the Lipschitz constant). References to some algorithms from each group are reported in Table 2.1 where in the case of the differentiable objective function (2.2), (2.4) the geometric methods are further differentiated by the type of the used minorant function that can be either non-smooth or smooth. Notice that methods working with smooth minorants and using multiple estimates of the Lipschitz constants for derivatives have not been proposed yet. The choice of the algorithms is explained mainly by the following two aspects. First, they are sufficiently representative for demonstrating various approaches used in the literature to estimate the Lipschitz constant. Second, they manifest a good performance on various sets of test and practical functions from the literature, and, therefore, are often chosen as worthy candidates for multidimensional extensions.

2.2 Geometric LGO Methods for Non-differentiable Functions

One of the first and well studied methods for solving the one-dimensional LGO problem (2.2), (2.3) is the already mentioned Piyavskij–Shubert method (see [245, 246, 306]), discussed in a vast literature (see, e.g., the survey [140] and references given therein). Various modifications of this algorithm have been also proposed (see, e.g., [22, 82, 140, 150, 266, 304, 328, 333]). We start this section from a detailed description of this algorithm due to its methodological importance for the further analysis of geometric LGO approaches.

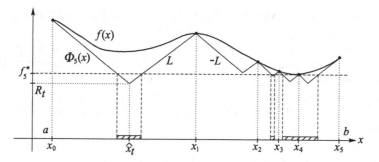

Fig. 2.3 The Lipschitzian objective function $f(x)$ (with the Lipschitz constant L) and its minorant $\Phi_5(x)$; sub-intervals which can contain the global minimizer of $f(x)$ during the work of the
Piyavskij–Shubert algorithm are drawn by hatching

The Piyavskij–Shubert method is a sequential algorithm that uses a priori given
information about the Lipschitz constant (namely, a given value of the Lipschitz
constant or its overestimate) to adaptively construct minorant auxiliary functions
for the objective function $f(x)$ as mentioned in the introductory Chap. 1. At the
first iteration $k = 1$ of the algorithm, initial trials are performed at several points
$x_0, x_1, \ldots, x_{n(1)}$ (usually, the left and right margins of the search interval $[a, b]$ are
chosen for this scope with $n(1) = 1$; hereinafter, only the variant $n(1) = 1$ will be
considered). At each successive iteration $k > 1$, trial points x_i (in the order of growth
of their coordinates) and the corresponding function values $z_i = f(x_i)$, $0 \leq i \leq k$,
are taken and the current lower bounding function $\Phi_k(x)$ is constructed as the union
of minorants $\phi_i(x)$ from (2.5) over all sub-intervals $[x_{i-1}, x_i]$, $1 \leq i \leq k$. Then, a
sub-interval with the minimal characteristic R_t is determined (see Fig. 2.3) taking
into account (2.6). Finally, a new trial point $\hat{x}_t$ is calculated on this sub-interval
by formula (2.7). In such a way, a new information about the objective function
is acquired by evaluating $f(\hat{x}_t)$, the piecewise linear minorant function is updated,
whereas the value

$$x_k^* = \arg \min_{0 \leq i \leq k} z_i$$

is chosen as a new approximation of the global minimizer from (2.2), with

$$f_k^* = \min_{0 \leq i \leq k} z_i$$

being the current approximation of the global minimum f^*. The trial points are
reordered and this iterative process is repeated until some stopping criterion is verified
(e.g., until the sub-interval with a new trial point becomes sufficiently small).

   In Fig. 2.3, an example of the lower bounding function $\Phi_k(x)$ constructed after six
function trials (i.e., after $k = 5$ iterations) is shown by continuous thin line. The black
dots on the objective function graph (thick line) indicate function values at the ordered
trial points $x_0, x_1, \ldots, x_5$. The next, i.e., the seventh, trial will be performed at the

point $\hat{x}_t = \hat{x}_1$ (see (2.7)). After executing this new trial, the lower bounding function will be reconstructed (precisely, $\Phi_6(x)$ will differ from $\Phi_5(x)$ over the sub-interval $[x_0, x_1]$ in Fig. 2.3), thus making better the current approximation of the problem solution. Note that some sub-intervals (namely, those for which the characteristic value is greater than the estimate f_k^* of the global minimum value; see, for instance, the sub-interval $[x_1, x_2]$ in Fig. 2.3) can be eliminated in order to reduce the region where x^* can be located. More precisely, the global minimizer x^* can be found only within the set $X^*(k)$ defined as

$$X^*(k) = \{x \in [a, b] : \Phi_k(x) \le f_k^*\}.$$

In Fig. 2.3, the sub-intervals forming the set $X^*(k)$ are indicated by hatching drawn over the axis x.

As observed, e.g., in [63, 152, 326], this algorithm is *optimal in one step*, i.e., the choice of the current evaluation point ensures the maximal improvement of the lower bound of the global minimum value of $f(x)$ with respect to a number of reasonable criteria (see, e.g., [150, 323, 326, 348] for the related discussions on optimality principles in Lipschitz global optimization). As many other geometric LGO methods, the Piyavskij–Shubert algorithm can be viewed as a branch-and-bound algorithm (see, e.g., [145, 150]) or, more generally, as a *divide-the-best algorithm* studied in [277]. Within this theoretical framework, the convergence of the sequence of trial points generated by the method to the global minimizers only (*global convergence*) can be easily established (see, e.g., [246, 277, 290]).

It should be noticed that branch-and-bound ideas were used also in the powerful method of non-uniform coverings proposed in [81, 82] (see also [87]) for functions with a priori given Lipschitz constants (see, e.g., [86, 88, 249] for modern high-performance realizations of this approach).

An interesting variation of the Piyavskij–Shubert algorithm has been proposed in [154] where the DIRECT method has been introduced. This algorithm iteratively selects several sub-intervals of the search region $[a, b]$ for partitioning and subdivides each of them into thirds with subsequent evaluations of the objective function $f(x)$ at the central points of new sub-intervals. The selection procedure is based on estimates of the lower bounds of $f(x)$ over sub-intervals by using a set of possible values for the Lipschitz constant, going from zero to infinity. In terms of the geometric approach, it is possible to say that all admissible minorant functions (in this case, they are piecewise linear discontinuous functions) are examined during the current iteration of the algorithm without constructing a specific one.

In Fig. 2.4, an example of a partition of $[a, b]$ into 5 sub-intervals performed by the DIRECT method is represented. The objective function $f(x)$ has been evaluated at the central points c_1, c_2, c_3, c_4, and c_5 of the corresponding sub-intervals (trial points are indicated by black dots in Fig. 2.4). Notice that at the first iteration the interval $[a, b]$ has been partitioned into three equal sub-intervals and the first three trials have been executed at the points c_3, c_1, and c_5. Then, the central sub-interval among the three ones has been divided into thirds and two new trials have been performed at the central points c_2 and c_4 of the left and the right of these thirds, respectively.

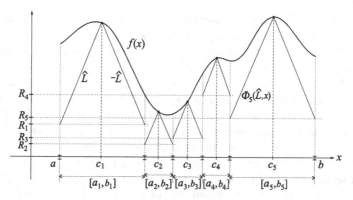

Fig. 2.4 Lower bounds R_i of $f(x)$ over sub-intervals $[a_i, b_i]$ corresponding to a particular estimate $\hat{L}$ of the Lipschitz constant

Given an overestimate $\hat{L}$ of the Lipschitz constant from (2.3), the objective function is bounded from below over $[a, b]$ by a piecewise linear discontinuous minorant function $\Phi_5(\hat{L}, x)$ (see Fig. 2.4; on the vertical axis the sub-intervals characteristics R_i, $1 \leq i \leq 5$, are indicated).

For the known overestimate $\hat{L}$ of the Lipschitz constant, new trials should be executed (similarly to the Piyavskij–Shubert algorithm) within the sub-interval $[a_2, b_2]$ having the minimal characteristic R_2 in order to obtain an improvement of the current estimate of the global minimum. But since in practical applications the exact Lipschitz constant L (or its overestimate) is often unknown, it is not possible to indicate with certainty that exactly this sub-interval is the most promising and it should be subdivided. In fact, the sub-interval $[a_2, b_2]$ in Fig. 2.4 has the smallest lower bound of $f(x)$ with respect to the estimate $\hat{L}$ of the Lipschitz constant. However, if a higher estimate $\tilde{L} \gg \hat{L}$ is taken, the slope of the lines in Fig. 2.4 increases and the sub-interval $[a_1, b_1]$ becomes preferable to all others since the lower bound of $f(x)$ over this sub-interval becomes the smallest one with respect to this new estimate $\tilde{L}$.

Jones et al. have proposed in [154] to use various estimates of the Lipschitz constant from zero to infinity at each iteration of their DIRECT algorithm. This corresponds to examination of all possible slopes $\hat{L}$ when auxiliary functions are considered (which in this case are not always minorants for $f(x)$) and lower bounds are calculated. Such a consideration leads to the basic idea of the DIRECT: to select for partitioning and sampling (i.e., performing the function trials) the so-called *potentially optimal sub-intervals*, i.e., sub-intervals over which $f(x)$ could have the best improvement with respect to a particular estimate of the Lipschitz constant. This is done by representing each sub-interval $[a_i, b_i]$ of the current partition of $[a, b]$ as a dot in a two-dimensional diagram with horizontal coordinate $c_i = (b_i - a_i)/2$ and vertical coordinate $f(c_i)$. It is then possible to prove that dots representing some potentially optimal sub-intervals are located on the lower right convex hull of all the dots corresponding to the sub-intervals (see [154]).

Thus, the DIRECT method is essentially the Piyavskij–Shubert algorithm modified to use a center-sampling strategy and to subdivide all potentially optimal subintervals related to different estimates of the Lipschitz constant L. Since during the search it uses a set of possible estimates of L and does not use its single overestimate, only the so-called *everywhere dense convergence* (i.e., convergence of the sequence of trial points to any point of the search interval) can be established for this method. It should be also noticed that it is difficult to apply for the DIRECT some meaningful stopping criterion, such as, for example, stopping on achieving a desired accuracy in solution. Nevertheless, due to its relative simplicity and a satisfactory performance on several test functions and applied problems, the DIRECT has been widely adopted in practical applications (see, e.g., [21, 30, 44, 57, 104, 141, 215, 338]) and has attracted the attention of many researchers (for its theoretical and experimental analysis and several modifications see, e.g., [53, 57, 93, 107, 128, 141, 154, 179, 189, 194, 197, 232, 289]).

As already observed, an assumption that the objective function satisfies the Lipschitz condition raises a question of estimating the corresponding Lipschitz constant. Strongin's algorithm (see, e.g., [315, 323]) answers this question by an adaptive estimation of the Lipschitz constant during the global search. It has a good convergence rate as compared to a number of other global optimization methods using only values of the objective function during the search (see, e.g., [132, 315, 323]). Therefore, this method has been often chosen as a good candidate for multidimensional extensions (see, e.g., [102, 176, 186, 216, 242, 286, 290, 323]).

Formally, this algorithm belongs to the class of the so-called *information-statistical algorithms* (or, just, *information algorithms*). The information approach originated in the works [220, 313] (see also [315, 323]) and, together with the Piyavskij–Shubert algorithm and non-uniform covering methods, it has consolidated foundations of the Lipschitz global optimization. The main idea of this approach is to apply the theory of random functions to building a mathematical representation of an available (certain or uncertain) a priori information on the objective function behavior. A systematic approach to the description of some uncertain information on this behavior is to accept that the unknown black-box function to be minimized is a sample of some known random function. Generally, to provide an efficient analytical technique for deriving some estimates of the global optimum with a finite number of trials, i.e., for obtaining by Bayesian reasoning some conditional (with respect to the trials performed) estimations, the random function should have some special structure. It is then possible to deduce the decision rules for performing new trials as some optimal decision functions (see, e.g., [29, 138, 155, 210, 212, 315, 323, 329, 348–350, 356, 357]).

In the Strongin algorithm, a global estimate H^k of the Lipschitz constant L is adaptively calculated by using the obtained results $z_i = f(x_i)$ of the function trials at the ordered trial points $x_i, 0 \leq i \leq k$,

$$H^k = H(k) = r \cdot \max_{1 \leq i \leq k} \frac{|z_i - z_{i-1}|}{x_i - x_{i-1}}, \tag{2.8}$$

where $r > 1$ is the algorithm parameter. It is possible to prove that for each fixed Lipschitzian function $f(x)$ there exists a value r^* such that taking $r \geq r^*$ guarantees convergence of the sequence of trial points generated by the method to the global minimizers of $f(x)$ only (see [315, 323]).

The Strongin algorithm can be considered in the framework of divide-the-best algorithms (see [277]). As demonstrated, e.g., in [216, 323], there is a firm relation between the information and geometric approaches. In fact, the characteristics R_i of the Strongin information algorithm associated with each sub-interval $[x_{i-1}, x_i]$, $1 \leq i \leq k$, of the search interval $[a, b]$ (see Fig. 2.2) can be rewritten (see [216, 323]) in a form similar to that of the Piyavskij–Shubert algorithm. This interesting fact allows one to interpret the Strongin method as a geometric algorithm adaptively constructing auxiliary piecewise linear functions during its work.

The usage of the global information only about behavior of the objective function, like the just mentioned global estimates of the Lipschitz constant, can lead to a slow convergence of algorithms to global minimizers. One of the traditional ways of overcoming this difficulty (see, e.g., [11, 148, 150, 194] and references given therein) recommends stopping the global procedure and switching to a local minimization method in order to improve the current solution and to accelerate the search during its final phase. Applying this technique can result in some problems related to the combination of global and local phases. The main problem is determining the moment to stop the global procedure and to start the local one. A premature arrest can provoke the loss of the global solution whereas a late one can slow down the search.

For example, it is well known that the DIRECT method balances global and local information during its work. However, the local phase is too pronounced in this balancing. The DIRECT executes too many function trials in attraction regions of local optima and, therefore, manifests a slow convergence to the global minimizers when the objective function has many local minimizers. In [289], a new geometric algorithm inspired by the DIRECT ideas has been proposed to solve difficult multiextremal LGO problems. To accomplish this task, a two-phase approach consisting of explicitly defined global and local phases has been incorporated in this method providing so a faster convergence to the global minimizers. Another type of local improvement strategy has been introduced in [186, 187, 301, 303]. This technique forces the global optimization method to make a local improvement of the best approximation of the global minimum immediately after a new approximation better than the current one is found.

In [272, 273] (see also [290, 301, 323]), another fruitful approach (the so-called *local tuning approach*) which allows global optimization algorithms to tune their behavior to the shape of the objective function at different sub-intervals has been proposed. The main idea behind this approach lies in the adaptive balancing of local and global information obtained during the search for every sub-interval $[x_{i-1}, x_i]$, $1 \leq i \leq k$, formed by trial points x_i. When a sub-interval $[x_{i-1}, x_i]$ is narrow, the local information obtained within the near vicinity of the trial points x_{i-1} and x_i has the major influence on the method behavior. In this case, the results of trials executed at points lying far from the interval $[x_{i-1}, x_i]$ are less significant for the method. In contrast, working with a wide sub-interval the method takes into consideration the global search information obtained from the whole search interval.

Both the comparison and the balancing of global and local data are effected by estimating local Lipschitz constants l_i for each sub-interval $[x_{i-1}, x_i], 1 < i \leq k$, as follows:

$$l_i = r \cdot \max\{\lambda_i, \gamma_i, \xi\}, \tag{2.9}$$

where

$$\lambda_i = \max\{H_{i-1}, H_i, H_{i+1}\}, \quad i = 2, \ldots, k,$$

$$H_i = \frac{|z_i - z_{i-1}|}{x_i - x_{i-1}}, \quad i = 2, \ldots, k, \tag{2.10}$$

$$H^k = \max\{H_i : i = 2, \ldots, k\}. \tag{2.11}$$

Here, $z_i = f(x_i), i = 1, \ldots, k$, i.e., values of the objective function calculated at the previous iterations at the trial points $x_i, i = 1, \ldots, k$, (when $i = 2$ and $i = k$ only H_2, H_3, and H_{k-1}, H_k, should be considered, respectively). The value γ_i is calculated as follows:

$$\gamma_i = H^k \frac{(x_i - x_{i-1})}{X^{max}}, \tag{2.12}$$

with H^k from (2.11) and

$$X^{max} = \max\{x_i - x_{i-1} : i = 2, \ldots, k\}. \tag{2.13}$$

The value H^k is an estimate of the global Lipschitz constant L over the interval $[a, b]$. The estimate l_i of the local Lipschitz constant L_i over an interval $[x_{i-1}, x_i]$ contains the following two fundamental parts: λ_i, which accounts for local properties, and γ_i, which accounts for global ones. If the interval $[x_{i-1}, x_i]$ is large, the global part increases, because in this case local information may not be reliable. In the opposite case, for small intervals $[x_{i-1}, x_i]$, the global part decreases, because local information is of the major importance and the global one loses its influence. Thus, at every sub-interval a balancing of local and global information is performed automatically.

Similarly to the Piyavskij–Shubert method, this algorithm constructs auxiliary functions approximating the objective function. Although these functions are not always minorants for $f(x)$ over the whole search interval, they are iteratively improved during the search in order to obtain appropriate bounds for the global minimum from (2.2). In Fig. 2.5, an example of the auxiliary function $\hat{\Phi}_k(x)$ for a

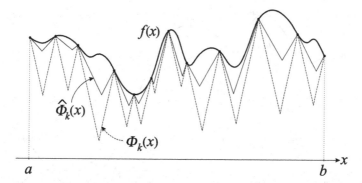

Fig. 2.5 An auxiliary function $\hat{\Phi}_k(x)$ (*solid thin line*) and a lower bounding function $\Phi_k(x)$ (*dashed line*) for a Lipschitz function $f(x)$ over $[a, b]$, constructed by using local Lipschitz estimates and by using the global Lipschitz constant, respectively

Lipschitz function $f(x)$ over $[a, b]$ constructed by using estimates of local Lipschitz constants over sub-intervals of $[a, b]$ is shown by a solid thin line; a lower bounding function $\Phi_k(x)$ for $f(x)$ over $[a, b]$ constructed by using an overestimate of the global Lipschitz constant is represented by a dashed line. It can be seen that $\hat{\Phi}_k(x)$ estimates the behavior of $f(x)$ over $[a, b]$ more accurately than $\Phi_k(x)$, especially over sub-intervals where the corresponding local Lipschitz constants are significantly smaller than the global one.

The local tuning approach enjoys the following properties (see [272, 273, 323]):

(1) the problem of determining the moment when to stop the global procedure does not arise because the local information is taken into consideration throughout the whole duration of the global search;

(2) the local information is taken into account not only in the neighborhood of a global minimizer but also over the whole search interval, thus allowing an additional acceleration of the global search;

(3) in order to guarantee convergence to the global minimizer x^* from (2.2) it is not necessary to know the exact Lipschitz constant over the whole search interval; on the contrary, only an overestimate of a local Lipschitz constant at a neighborhood of x^* is needed;

(4) geometric local tuning algorithms can be successfully parallelized (see, e.g., [284–286, 323]) and easily extended to the multidimensional case (see, e.g., [176, 186, 272, 290, 292, 303, 323]).

These advantages allow one to adopt the local tuning approach for an efficient solving different univariate and multidimensional LGO problems (see, e.g., [175, 186, 216, 272, 273, 276, 278, 290, 292, 303, 323]).

To conclude our presentation of the Lipschitz geometric ideas for solving the LGO problem (2.2)–(2.3), let us return to the problem of finding the minimal root of an equation with multiextremal non-differentiable left part (see (2.1) and Fig. 2.1 in Chap. 1). A possible fast and efficient algorithm for solving this important practical

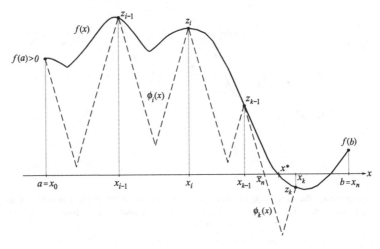

Fig. 2.6 Finding the minimal root of equation $f(x) = 0$ with multiextremal non-differentiable *left* part by a geometric method

problem can be developed as follows (see, e.g., [290, 323]). Let us suppose that the objective function $f(x)$ has been already evaluated at some trial points x_i, $0 \leq i \leq n$, and $z_i = f(x_i)$ (see Fig. 2.6). For every interval $[x_{i-1}, x_i]$, $1 \leq i \leq n$, a piecewise linear function $\phi_i(x)$ is constructed (it is drawn by dashed line in Fig. 2.6) by using the Lipschitz information in such a way that $\phi_i(x) \leq f(x)$, $x \in [x_{i-1}, x_i]$. By knowing the structure of the auxiliary function $\phi_i(x)$, $1 \leq i \leq n$, it is possible to determine the minimal index $k \geq 1$, such that the equation $\phi_k(x) = 0$ has a solution (point $\tilde{x}_n$ in Fig. 2.6) over $[x_{k-1}, x_k]$. Adaptively improving the set of functions $\phi_i(x)$, $1 \leq i \leq n$, by adding new trial points $\tilde{x}_n$, $n > 1$, we improve both our lower approximation of $f(x)$ and the current solution to the problem. In this manner, such a geometric method either finds the minimal root of the equation $f(x) = 0$ or determines the global minimizer of $f(x)$ (in the case when the equation under consideration has no roots on the given interval). Its performance is significantly faster in comparison with the methods traditionally used by engineers for solving this problem. The usage of the local tuning technique on the behavior of $f(x)$ allows one to obtain a further acceleration of the search (see, e.g., [218, 290, 323]).

2.3 Geometric LGO Methods for Differentiable Functions with the Lipschitz First Derivatives

The restriction of the class of the objective functions (i.e., the examination of the LGO problem (2.2), (2.4) rather than the problem (2.2), (2.3)) opens new opportunities for developing efficient geometric LGO methods. In fact, if at each point $x \in [a, b]$ it is possible to evaluate both the objective function $f(x)$ and its first derivative $f'(x)$,

Fig. 2.7 Non-smooth piecewise quadratic auxiliary function $\psi_i(x)$ which can be constructed over a sub-interval $[x_{i-1}, x_i]$ for a function $f(x)$ with the Lipschitz first derivative

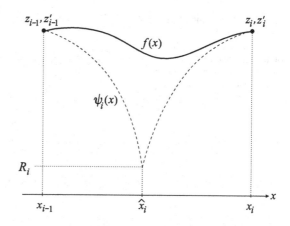

this gives the opportunity to obtain more information about the problem (especially, regarding its local properties represented by the derivative values). The usage of this information allows one to construct auxiliary functions that fit closely the objective function and, therefore, to accelerate the global search.

The geometric approach for solving the LGO problem (2.2), (2.4) has obtained a strong impact to expansion after publication of the papers [34, 116]. In these articles, non-smooth piecewise quadratic minorants have been used to approximate the behavior of the objective function $f(x)$ from (2.2) by using the Lipschitz condition (2.4).

In Fig. 2.7, an example of such a lower bounding function $\psi_i(x)$ over a sub-interval $[x_{i-1}, x_i]$ is shown. Both the objective function $f(x)$ from (2.2) and its first derivative $f'(x)$ satisfying the Lipschitz from (2.4) have been evaluated at two points x_{i-1} and x_i of the search interval $[a, b]$ with the corresponding values $z_{i-1} = f(x_{i-1})$, $z'_{i-1} = f'(x_{i-1})$ and $z_i = f(x_i)$, $z'_i = f'(x_i)$. If an overestimate $m \geq K$ is known, the lower bounding function $\psi_i(x)$ for $f(x)$ over $[x_{i-1}, x_i]$ can be constructed using the Lipschitz condition (2.4) and the Taylor formula for $f(x)$ as follows:

$$\psi_i(x) = \max \{ z_{i-1} + z'_{i-1}(x - x_{i-1}) - 0.5m(x - x_{i-1})^2, \\ z_i - z'_i(x_i - x) - 0.5m(x_i - x)^2 \}. \tag{2.14}$$

The lower bound value R_i for $f(x)$ over $[x_{i-1}, x_i]$ (the sub-interval characteristic) is shown in Fig. 2.7. It can be explicitly calculated in a way similar to (2.6) and, as a rule, it results to be closer to the minimal value of $f(x)$ over $[x_{i-1}, x_i]$ than the piecewise linear minorants (2.5). This can accelerate the global search being a natural consequence of the availability and the usage of more complete information about the function in the LGO problem (2.2), (2.4) with respect to the problem (2.2), (2.3).

As in the case of the LGO problem (2.2), (2.3), various approaches can be used to obtain an estimate m of the Lipschitz constant K for constructing auxiliary functions over the whole interval $[a, b]$ as the union of functions $\psi_i(x)$ from (2.14). For example, in [34] (see also [14, 15]), the constant K from (2.4) is supposed to be a

Fig. 2.8 Smooth piecewise quadratic auxiliary function $\theta_i(x)$ (*dashed line*) for the objective function $f(x)$ (*thick line*) with the Lipschitz first derivative over $[x_{i-1}, x_i]$

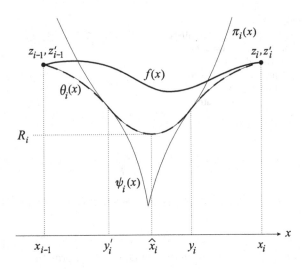

priori known. In [116] (see also [323]), an adaptive global estimate m of the constant K during the function minimization is proposed.

In [270, 275], it was shown that the Lipschitz constant K can be estimated more accurately over the whole interval $[a, b]$ in comparison with [116] and that the local tuning approach can be used in a similar to (2.9) manner, thus providing a significant acceleration of the global search (see also [290, 323]). It is important to emphasize that in order to ensure the convergence to the global minimizer x^* from (2.2), it is not necessary to estimate correctly the global Lipschitz constant K (it may be underestimated) during the execution of the local tuning algorithm but it is sufficient to have an overestimate only of the local Lipschitz constant over a sub-interval containing the point x^*.

It is evident from Fig. 2.7 that the auxiliary functions based on (2.14) are not smooth at the points $\hat{x}_i$ in spite of the smoothness of the objective function $f(x)$ over $[a, b]$. In [271, 276], it has been demonstrated how to obtain smooth auxiliary functions making them closer to $f(x)$ than those shown in Fig. 2.7 and, therefore, accelerating the global search (see also [15, 119, 188, 200, 320] where similar constructions are discussed). A general scheme describing the methods using smooth bounding procedures has been introduced in [271, 276] with several approaches for the Lipschitz constant estimation (a priori given, global, and local estimates were considered). Let us present this scheme.

The construction of a smooth auxiliary function over $[x_{i-1}, x_i]$ is based on the following considerations. The objective function $f(x)$ should be strictly above the function $\psi_i(x)$ for all $x \in (y_i', y_i)$ (see Fig. 2.8) because due to (2.4) its curvature is bounded by a parabola

$$\pi_i(x) = 0.5mx^2 + b_i x + c_i,$$

where the unknowns y_i', y_i, b_i, and c_i can be determined by solving the following system of equations:

$$\begin{cases} \psi_i(y_i') = \pi_i(y_i'), \\ \psi_i(y_i) = \pi_i(y_i), \\ \psi_i'(y_i') = \pi_i'(y_i'), \\ \psi_i'(y_i) = \pi_i'(y_i). \end{cases}$$

Here the first equation provides the coincidence of $\psi_i(x)$ and $\pi_i(x)$ at the point y_i' and the third one provides the coincidence of their derivatives $\psi_i'(x)$ and $\pi_i'(x)$ at the same point. The second and fourth equations provide the fulfillment of analogous conditions at the point y_i.

Thus, once the values y_i', y_i, b_i, and c_i are determined (see [271, 276, 290, 323] for detailed discussions describing how to get explicit formulae for them), it may be concluded that the following function

$$\theta_i(x) = \begin{cases} \psi_i(x), \ x \in [x_{i-1}, y_i') \cup [y_i, x_i], \\ \pi_i(x), \ x \in [y_i', y_i), \end{cases}$$

is a smooth piecewise quadratic auxiliary function for $f(x)$ over $[x_{i-1}, x_i]$ (see Fig. 2.8), i.e., that the first derivative $\theta_i'(x)$ exists over $x \in [x_{i-1}, x_i]$ and

$$f(x) \geq \theta_i(x), \quad x \in [x_{i-1}, x_i].$$

As demonstrated also by extensive numerical experiments (see, e.g., [45, 276, 290, 295, 323]), the performance of geometric methods for solving the LGO problem (2.2), (2.4) with smooth auxiliary functions overcomes that of geometric methods with non-smooth minorants. The usage of the local tuning technique, in its turn, ensures the further speed up of the methods, especially when a high accuracy of the problem solution is required.

It is worthy to mention that geometric methods based on construction of the Lipschitz piecewise quadratic auxiliary functions can be also applied to solving efficiently the problem of finding the minimal root of an equation with multiextremal differentiable left part, discussed above (see, e.g., [158, 282, 290, 323]).

Up to now, geometric methods for solving the LGO problem (2.2), (2.4) that use in their work an a priori given estimate of K from (2.4), its adaptive global estimate or adaptive estimates of local Lipschitz constants have been considered. Algorithms working with multiple estimates of the Lipschitz constant for $f'(x)$ chosen from a set of possible values were not known until 2009 (see [177]) in spite of the fact that a geometric method working in this way with the Lipschitz objective functions (the DIRECT method described in the previous section) has been proposed in 1993 (see [154]). The main obstacle in implementing such an algorithm for differentiable objective functions was a lack of an efficient procedure for determining sub-intervals to perform new trials. A geometric method resolving this problem in a simple way and evolving the DIRECT ideas to the case of the objective function having the Lipschitz first derivative has been introduced and studied in [177].

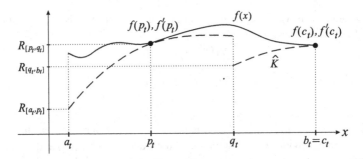

Fig. 2.9 Subdivision of a sub-interval $[a_t, b_t]$ in the situation where $f(x)$ and $f'(x)$ are evaluated at the point p_t given that they have been previously evaluated at the point b_t; a discontinuous piecewise quadratic auxiliary function for $f(x)$ is drawn by a *dashed line*

In this algorithm, the partition of the search interval $[a, b]$ into sub-intervals $[a_i, b_i]$ is iteratively performed by subdividing a selected sub-interval $[a_t, b_t]$ (see Fig. 2.9) into three equal parts of the length $(b_t - a_t)/3$, i.e.,

$$[a_t, b_t] = [a_t, p_t] \cup [p_t, q_t] \cup [q_t, b_t],$$
$$p_t = a_t + (b_t - a_t)/3, \quad q_t = b_t - (b_t - a_t)/3.$$

A new trial is carried out either at the point p_t (if both the objective function $f(x)$ and its first derivative $f'(x)$ have been evaluated over the sub-interval $[a_t, b_t]$ at the point b_t, see Fig. 2.9), or at the point q_t (if $f(x)$ and $f'(x)$ have been evaluated at the point a_t). Notice that an efficient partition strategy is adopted here since each selected sub-interval is subdivided by only one new trial point into three new sub-intervals. Remind that in a center-sampling partition strategy, in order to have three new sub-intervals, it is necessary to perform the function evaluations at two new points.

A series of non-smooth (discontinuous) piecewise quadratic functions corresponding to different estimates $\hat{K}$ of the Lipschitz constants K from (2.4) is then taken into account to find approximations $R_i = R_{[a_i, b_i]}$ of the lower bounds of $f(x)$ over sub-intervals $[a_i, b_i]$ (see Fig. 2.9); given an estimate $\hat{K}$, a lower bound R_i of the function values over the sub-interval $[a_i, b_i]$ can be calculated as

$$R_i = f(c_i) \pm f'(c_i)(b_i - a_i) - 0.5\hat{K}(b_i - a_i)^2, \tag{2.15}$$

where the sign '$-$' is used in the case of the right-end function evaluation, i.e., $c_i = b_i$, and the sign '$+$' is used in the case of the left-end function evaluation, i.e., $c_i = a_i$ (see Fig. 2.9).

In the DIRECT method for solving the problem (2.2), (2.3), the potentially optimal sub-intervals are determined as candidates for partitioning at each iteration by varying estimates $\hat{L}$ of the Lipschitz constant L from zero to infinity. The determination of these sub-intervals is a relatively simple technical task. In fact, it is sufficient

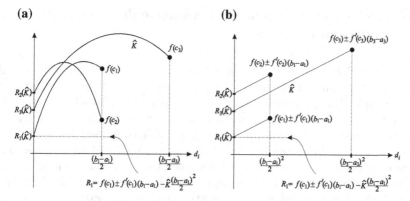

Fig. 2.10 Different ways of a graphical representation of sub-intervals in geometric methods for solving the LGO problem (2.2),(2.4) based on multiple estimates of the Lipschitz constant (2.4) with non-smooth piecewise quadratic auxiliary functions

to represent each sub-interval $[a_i, b_i]$ as a dot in a two-dimensional diagram with horizontal coordinate $c_i = (b_i - a_i)/2$ and vertical coordinate $f(c_i)$ and to locate the dots on a lower right convex hull of all the dots.

When solving the problem (2.2),(2.4), the same operation gives troubles due to the nonlinear part in (2.15). The difficulties in establishing the relation of domination (in terms of the lower bounds R_i) between sub-intervals of a current partition generated by the method are illustrated in Fig. 2.10a. Here, the sub-intervals $[a_1, b_1]$ and $[a_3, b_3]$ are the so-called *non-dominated sub-intervals* (i.e., sub-intervals having the smallest lower bound for some particular estimate of the Lipschitz constant for $f'(x)$; see [177] for details); but it is impossible to determine this fact from the diagram in Fig. 2.10a.

An efficient solution to this inconvenience is given by a new diagram in Fig. 2.10b where the intersection of the line with a slope $\hat{K}$ passed through any dot representing a sub-interval and the vertical coordinate axis gives us the lower bound (2.15) of $f(x)$ over the corresponding sub-interval. Thus, the procedure of selecting sub-intervals to be partitioned becomes simple in the case of differentiable objective functions too (the two non-dominated sub-intervals have in Fig. 2.10b coordinates $(0.5(b_1 - a_1), f(c_1) \pm f'(c_1)(b_1 - c_1))$ and $(0.5(b_3 - a_3), f(c_3) \pm f'(c_3)(b_3 - c_3))$, respectively) and another practical geometric method for solving the LGO problem (2.2), (2.4) is also available (see [177]). The usage of derivatives allows one to obtain, as it is expected, an acceleration in comparison with the DIRECT method.

Notice that the development of a geometric method constructing smooth auxiliary functions with multiple estimates of the Lipschitz constant (2.4) for derivatives still remains an open problem.

2.4 Acceleration Techniques Embedded in the Univariate Global Optimization

Very often in global optimization local techniques are used to accelerate the global search, and frequently global and local searches are realized by different methods having completely alien structures. Such a combination introduces at least two inconveniences. First, function trials executed by a local search procedure are not used typically in the subsequent phases of the global search. Only some results of these trials (for instance, the current best found value) can be used, and the other ones are not taken into consideration by the global search method. Second, there arises the necessity to introduce both a rule that stops the global phase and starts the local one, and a rule that stops the local phase and decides whether it is necessary to re-start the global search. Clearly, a premature stop of the global phase of the search can lead to the loss of the global solution, while a late stop of the global phase can slow down the search.

In this section, both frameworks, geometric and information, are taken into consideration and several derivative-free techniques proposed to accelerate the global search are studied and compared. Some promising ideas that can be used to speed up the search both in the framework of geometric and information algorithms are introduced. All the acceleration techniques have the advantage to get over both the difficulties mentioned above, namely:

- the accelerated global optimization methods automatically realize a local behavior in the promising subregions without the necessity to stop the global optimization procedure;
- all the trials executed during the local phases are used in the course of the global phases, as well.

It should be emphasized that the resulting geometric and information global optimization methods have a similar structure, and a smart mixture of new and traditional computational steps leads to different global optimization algorithms. All of them are studied and compared on three sets of tests: the widely used set of 20 test functions taken from [140]; 100 randomly generated functions from [243]; and four functions arising in practical problems [120, 290].

As mentioned in the previous sections (see formulae (2.9)–(2.13)), the *local tuning technique* proposed in [272, 273] adaptively estimates *local* Lipschitz constants at different sub-intervals of the search region during the course of the optimization process. The two components, λ_i and γ_i, are the main players in (2.9). They take into account, respectively, the local and the global information obtained during the previous iterations. When the interval $[x_{i-1}, x_i]$ is large, the local information represented by λ_i can be not reliable and the global part γ_i has a decisive influence on l_i thanks to (2.9) and (2.12). In this case $\gamma_i \to H^k$, namely, it tends to the estimate of the global Lipschitz constant L. In contrast, when $[x_{i-1}, x_i]$ is narrow, then the local information becomes relevant, the estimate γ_i becomes small (see (2.12)), and the local component λ_i assumes the key role. Thus, the local tuning technique automatically balances the global and the local information available at the current iteration. It has been proved for a number of global optimization algorithms that the usage of the local tuning can accelerate the search significantly (see [175, 273, 276, 281, 287, 290, 300, 323]). This local tuning strategy will be called *"Maximum" Local Tuning* hereinafter.

Recently, a new local tuning strategy called hereinafter *"Additive" Local Tuning* has been proposed in [115, 119, 320] for certain information algorithms. It proposes to use the following additive convolution instead of (2.9):

$$l_i = r \cdot \max\{\frac{1}{2}(\lambda_i + \gamma_i), \xi\}, \tag{2.16}$$

where r, ξ, λ_i, and γ_i have the same meaning as in (2.9). The first numerical examples executed in [115, 320] have shown a very promising performance of the "Additive" Local Tuning. These results induced us to execute in [301] a broad experimental testing and a theoretical analysis of the "Additive" Local Tuning. In particular, geometric methods using this technique were proposed in [301] (remind that the authors of [115, 320] have introduced (2.16) in the framework of information methods only). During our study some features suggesting a careful usage of this technique have been discovered, especially in cases where it is applied to geometric global optimization methods.

In order to start our analysis of the "Additive" Local Tuning, let us remind (see, e.g., [242, 246, 290, 303, 315, 323]) that in both the geometric and the information univariate algorithms, an interval $[x_{t-1}, x_t]$ is chosen in a certain way at the $(k + 1)$-th iteration of the optimization process and a new trial point, x^{k+1}, where the $(k + 1)$-th evaluation of $f(x)$ is executed, is computed as follows:

$$x^{k+1} = \frac{x_t + x_{t-1}}{2} - \frac{z_t - z_{t-1}}{2l_t}. \tag{2.17}$$

For a correct work of this kind of algorithms it is necessary that x^{k+1} is such that $x^{k+1} \in (x_{t-1}, x_t)$. It is easy to see that the necessary condition for this inclusion is $l_t > H_t$, where H_t is calculated as in (2.9). Thus, if the estimate l_t is obtained by using (2.16), then the sum of the two addends (λ_i and γ_i) plays the leading role. Since the estimate γ_i is calculated as shown in (2.12), it can be very small for small intervals, creating so the possibility of occurrence of the situation $l_t \leq H_t$, leading to $x^{k+1} \notin (x_{t-1}, x_t)$ that provokes an incorrect work of the algorithm using (2.16). Obviously, by increasing the value of the parameter r this unhappy situation can be easily avoided but the method should be re-started. In fact, in information algorithms where $r \geq 2$ is usually used, this risk is less pronounced, while in geometric methods where $r > 1$ is applied it becomes more probable. On the other hand, it is well known in Lipschitz global optimization (see, e.g., [242, 290, 303, 323]) that increasing the parameter r can slow down the search. In order to understand better the functioning of the "Additive" Local Tuning, it is broadly tested in the numerical part of this chapter together with other competitors (see Sect. 2.5).

The analysis provided above shows that the usage of the "Additive" Local Tuning can become tricky in some cases. In order to avoid the necessity to check the satisfaction of the condition $x^{k+1} \in (x_{t-1}, x_t)$ at each iteration, a new strategy called the "*Maximum-Additive*" *Local Tuning* has been proposed in [301] where, on the one hand, this condition is satisfied automatically and, on the other hand, advantages of both the local tuning techniques described above are incorporated in the unique strategy. This local tuning strategy calculates the estimate l_i of the local Lipschitz constants as follows:

$$l_i = r \cdot \max\{H_i, \frac{1}{2}(\lambda_i + \gamma_i), \xi\}, \tag{2.18}$$

where r, ξ, H_i, λ_i, and γ_i have the usual meaning (see (2.9)–(2.13)). It can be seen from (2.18) that this strategy both maintains the additive character of the convolution and satisfies condition $l_i > H_i$. The latter condition provides that in case the interval $[x_{i-1}, x_i]$ is chosen for subdivision (i.e., $t := i$ is assigned), the new trial point x^{k+1} will belong to the open interval (x_{t-1}, x_t). Notice that in (2.18) the equal usage of the local and global estimate is applied. Obviously, a more general scheme similar to (2.16) and (2.18) can be used, where $\frac{1}{2}$ is substituted by different weights for the estimates λ_i and γ_i, for example, as follows:

$$l_i = r \cdot \max\{H_i, \frac{\lambda_i}{r} + \frac{r-1}{r}\gamma_i, \xi\}$$

where $r, H_i, \lambda_i, \gamma_i$, and ξ are as in (2.18).

Let us now present another acceleration idea from [301]. It consists of the following observation, related to global optimization problems with a fixed budget of possible evaluations of the objective function $f(x)$, i.e., when only, for instance, 100 or 1 000 evaluations of $f(x)$ are allowed (see problem (P2) in the introductory Chap. 1). In these problems, it is necessary to obtain the best possible value of $f(x)$ as soon as possible. Suppose that f_k^* is the best value (the record value) obtained after k iterations. If a new value $f(x^{k+1}) < f_k^*$ has been obtained, then it can make sense to try to improve this value locally, instead of continuing the usual global search phase. As was already mentioned, traditional methods stop the global procedure and start a local descent: trials executed during this local phase are not then used by the global search since the local method has usually a completely different nature.

In order to overcome this drawback, two *local improvement techniques*, the "*optimistic*" and the "*pessimistic*" ones, that perform the local improvement within the global optimization scheme have been studied in [301]. The optimistic method alternates the local steps with the global ones and, if during the local descent a new promising local minimizer is not found, then the global method stops when a local stopping rule is satisfied. The pessimistic strategy does the same until the satisfaction of the required accuracy on the local phase, and then switches to the global phase where the trials performed during the local phase are also taken into consideration.

All the methods described in this section have a similar structure and belong to the class of divide-the-best global optimization algorithms introduced in [277] (see also [290]; for methods using the "Additive" Local Tuning this holds if the parameter r is such that $l_{i(k)} > r H_{i(k)}$ for all i and k). The algorithms differ in the following:

- methods are either geometric or information;
- methods differ in the way the Lipschitz information is used: an a priori estimate, a global estimate, and a local tuning;
- in cases where a local tuning is applied, methods use 3 different strategies: Maximum, Additive, and Maximum-Additive;
- in cases where a local improvement is applied, methods use either the optimistic or the pessimistic strategy.

Let us describe the General Scheme (GS) of the methods used in this section. A concrete algorithm will be obtained by specifying one of the possible implementations of Steps 2–4 in this GS.

Step 0. *Initialization.* Execute first two trials at the points a and b, i.e., $x^1 := a$, $z^1 := f(a)$ and $x^2 := b$, $z^2 := f(b)$. Set the iteration counter $k := 2$.
Let $flag$ be the local improvement switch to alternate global search and local improvement procedures; set its initial value $flag := 0$. Let $i_{\min}$ be an index (being constantly updated during the search) of the current record point, i.e., $z_{i_{\min}} = f(x_{i_{\min}}) \le f(x_i)$, $i = 1, \ldots, k$ (if the current minimal value is attained at several trial points, then the smallest index is accepted as $i_{\min}$).
Suppose that $k \ge 2$ iterations of the algorithm have already been executed. The iteration $k + 1$ consists of the following steps.
Step 1. *Reordering.* Reorder the points $x^1, \ldots, x^k$ (and the corresponding function values $z^1, \ldots, z^k$) of previous trials by subscripts so that

$$a = x_1 < \ldots < x_k = b, \quad z_i = f(x_i), \quad 1 \le i \le k.$$

Step 2. *Estimates of the Lipschitz constant.* Calculate the current estimates l_i of the Lipschitz constant for each sub-interval $[x_{i-1}, x_i]$, $i = 2, \ldots, k$, in one of the following ways.

Step 2.1. *A priori given estimate.* Take an a priori given estimate $\hat{L}$ of the Lipschitz constant for the whole interval $[a, b]$, i.e., set $l_i := \hat{L}$.
Step 2.2. *Global estimate.* Set $l_i := r \cdot \max\{H^k, \xi\}$, where r and ξ are two parameters with $r > 1$ and ξ sufficiently small, H^k is from (2.11).
Step 2.3. *"Maximum" Local Tuning.* Set l_i following (2.9).
Step 2.4. *"Additive" Local Tuning.* Set l_i following (2.16).
Step 2.5. *"Maximum-Additive" Local Tuning.* Set l_i following (2.18).

Step 3. *Calculation of characteristics.* Compute for each sub-interval $[x_{i-1}, x_i]$, $i = 2, \ldots, k$, its characteristic R_i by using one of the following rules.

Step 3.1. *Geometric methods.*

$$R_i = \frac{z_i + z_{i-1}}{2} - l_i \frac{x_i - x_{i-1}}{2}. \tag{2.19}$$

Step 3.2. *Information methods.*

$$R_i = 2(z_i + z_{i-1}) - l_i(x_i - x_{i-1}) - \frac{(z_i - z_{i-1})^2}{l_i(x_i - x_{i-1})}. \tag{2.20}$$

Step 4. *Sub-interval selection.* Determine sub-interval $[x_{t-1}, x_t]$, $t = t(k)$, for performing next trial by using one of the following rules.

  Step 4.1. *Global phase.* Select the sub-interval $[x_{t-1}, x_t]$ corresponding to the minimal characteristic, i.e., such that $t = \arg\min_{i=2,...,k} R_i$.
  Steps 4.2–4.3. *Local improvement.*
    if $flag = 1$, **then** *(perform local improvement)*
    ·   **if** $z^k = z_{i_{\min}}$, **then** $t = \arg\min\{R_i : i \in \{i_{\min} + 1, i_{\min}\}\}$;
    ·   **else** alternate the choice of sub-interval between $[x_{i_{\min}}, x_{i_{\min}+1}]$
        and $[x_{i_{\min}-1}, x_{i_{\min}}]$ starting from the right sub-interval $[x_{i_{\min}}, x_{i_{\min}+1}]$.
    ·   **end if**
    else $t = \arg\min_{i=2,...,k} R_i$ *(do not perform local improvement at the current iteration).*
    end if
The subsequent part of this Step differs for two local improvement techniques.
  Step 4.2. *Pessimistic local improvement.*

  ·   **if** $flag = 1$ and

$$x_t - x_{t-1} \leq \delta, \tag{2.21}$$

       where $\delta > 0$ is the local search accuracy,

  ·   **then** $t = \arg\min_{i=2,...,k} R_i$ *(local improvement is not performed since the local search accuracy has been achieved).*
  ·   **end if**
  ·   Set $flag := NOT(flag)$ *(switch the local/global flag).*
  Step 4.3. *Optimistic local improvement.*

  Set $flag := NOT(flag)$ *(switch the local/global flag: the accuracy of local search is not separately checked in this strategy).*

Step 5. *Global stopping criterion.* **If**

$$x_t - x_{t-1} \leq \varepsilon, \tag{2.22}$$

where $\varepsilon > 0$ is a given accuracy of the global search, **then Stop** and take as an estimate of the global minimum f^* the value $f_k^* = \min_{i=1,\ldots,k}\{z_i\}$ obtained at a point $x_k^* = \arg\min_{i=1,\ldots,k}\{z_i\}$.
Otherwise, go to Step 6.
Step 6. *New trial.* Execute next trial at the point x^{k+1} from (2.17): $z^{k+1} := f(x^{k+1})$. Increase the iteration counter $k := k + 1$, and go to Step 1.

All the Lipschitz global optimization methods considered in this section are summarized in Table 2.2, from which concrete implementations of Steps 2–4 in the GS can be individuated. As shown experimentally in the numerical part of this section, the methods using an a priori given estimate of the Lipschitz constant or its global estimate loss, as a rule, in comparison with the methods using local tuning techniques, in terms of the trials performed to approximate the global solutions to problems. Therefore, local improvement accelerations (Steps 4.2–4.3 of the GS) were implemented for methods using local tuning strategies only. In what follows, the methods from Table 2.2 are furthermore specified (for the methods known in the literature, the respective references are provided).

1. **Geom-AL**: Piyavskij–Shubert's method with the a priori given Lipschitz constant (see [246, 306] and [180, 290] for generalizations and discussions): GS with Step 2.1, Step 3.1, and Step 4.1.

2. **Geom-GL**: Geometric method with the global estimate of the Lipschitz constant (see [290]): GS with Step 2.2, Step 3.1, and Step 4.1.

3. **Geom-LTM**: Geometric method with the "Maximum" Local Tuning (see [273, 290, 323]): GS with Step 2.3, Step 3.1, and Step 4.1.

4. **Geom-LTA**: Geometric method with the "Additive" Local Tuning: GS with Step 2.4, Step 3.1, and Step 4.1.

5. **Geom-LTMA**: Geometric method with the "Maximum-Additive" Local Tuning: GS with Step 2.5, Step 3.1, and Step 4.1.

6. **Geom-LTIMP**: Geometric method with the "Maximum" Local Tuning and the pessimistic strategy of the local improvement (see [188, 290]): GS with Step 2.3, Step 3.1, and Step 4.2.

7. **Geom-LTIAP**: Geometric method with the "Additive" Local Tuning and the pessimistic strategy of the local improvement: GS with Step 2.4, Step 3.1, and Step 4.2.

8. **Geom-LTIMAP**: Geometric method with the "Maximum-Additive" Local Tuning and the pessimistic strategy of the local improvement: GS with Step 2.5, Step 3.1, and Step 4.2.

9. **Geom-LTIMO**: Geometric method with the "Maximum" Local Tuning and the optimistic strategy of the local improvement: GS with Step 2.3, Step 3.1, and Step 4.3.

10. **Geom-LTIAO**: Geometric method with the "Additive" Local Tuning and the optimistic strategy of the local improvement: GS with Step 2.4, Step 3.1, and Step 4.3.

Table 2.2 Description of the methods considered in this section, the signs "+" show a combination of implementations of Steps 2–4 in the GS for each method

Method	Step2					Step3		Step4		
	2.1	2.2	2.3	2.4	2.5	3.1	3.2	4.1	4.2	4.3
Geom-AL	+					+		+		
Geom-GL		+				+		+		
Geom-LTM			+			+		+		
Geom-LTA				+		+		+		
Geom-LTMA					+	+		+		
Geom-LTIMP			+			+			+	
Geom-LTIAP				+		+			+	
Geom-LTIMAP					+	+			+	
Geom-LTIMO			+			+				+
Geom-LTIAO				+		+				+
Geom-LTIMAO					+	+				+
Inf-AL	+						+	+		
Inf-GL		+					+	+		
Inf-LTM			+				+	+		
Inf-LTA				+			+	+		
Inf-LTMA					+		+	+		
Inf-LTIMP			+				+		+	
Inf-LTIAP				+			+		+	
Inf-LTIMAP					+		+		+	
Inf-LTIMO			+				+			+
Inf-LTIAO				+			+			+
Inf-LTIMAO					+		+			+

11. **Geom-LTIMAO**: Geometric method with the "Maximum-Additive" Local Tuning and the optimistic strategy of the local improvement: GS with Step 2.5, Step 3.1, and Step 4.3.

12. **Inf-AL**: Information method with the a priori given Lipschitz constant (see [290]): GS with Step 2.1, Step 3.2, and Step 4.1.

13. **Inf-GL**: Strongin's information-statistical method with the global estimate of the Lipschitz constant (see [314, 315, 323]): GS with Step 2.2, Step 3.2, and Step 4.1.

14. **Inf-LTM**: Information method with the "Maximum" Local Tuning (see [272, 303, 323]): GS with Step 2.3, Step 3.2, and Step 4.1.

15. **Inf-LTA**: Information method with the "Additive" Local Tuning (see [115, 320]): GS with Step 2.4, Step 3.2, and Step 4.1.

16. **Inf-LTMA**: Information method with the "Maximum-Additive" Local Tuning: GS with Step 2.5, Step 3.2, and Step 4.1.

17. **Inf-LTIMP**: Information method with the "Maximum" Local Tuning and the pessimistic strategy of the local improvement [186, 303]: GS with Step 2.3, Step 3.2, and Step 4.2.

18. **Inf-LTIAP**: Information method with the "Additive" Local Tuning and the pessimistic strategy of the local improvement: GS with Step 2.4, Step 3.2, and Step 4.2.

19. **Inf-LTIMAP**: Information method with the "Maximum-Additive" Local Tuning and the pessimistic strategy of the local improvement: GS with Step 2.5, Step 3.2, and Step 4.2.

20. **Inf-LTIMO**: Information method with the "Maximum" Local Tuning and the optimistic strategy of the local improvement: GS with Step 2.3, Step 3.2, and Step 4.3.

21. **Inf-LTIAO**: Information method with the "Additive" Local Tuning and the optimistic strategy of the local improvement: GS with Step 2.4, Step 3.2, and Step 4.3.

22. **Inf-LTIMAO**: Information method with the "Maximum-Additive" Local Tuning and the optimistic strategy of the local improvement: GS with Step 2.5, Step 3.2, and Step 4.3.

Let us spend a few words regarding convergence of the methods belonging to the GS. To do this, we study an infinite trial sequence $\{x^k\}$ generated by an algorithm belonging to the general scheme GS for solving the problem (2.2), (2.3) with $\delta = 0$ from (2.21) and $\varepsilon = 0$ from (2.22).

Theorem 2.1 *Assume that the objective function $f(x)$ satisfies the Lipschitz condition (2.3) with a finite constant $L > 0$, and let x' be any limit point of $\{x^k\}$ generated by an algorithm belonging to the GS, that does not use the "Additive" Local Tuning and works with one of the estimates (2.9), (2.11), (2.18). Then, the following assertions hold:*

1. *if $x' \in (a, b)$ then convergence to x' is bilateral, i.e., there exist two infinite subsequences of $\{x^k\}$ converging to x': one from the left, the other from the right;*
2. *$f(x^k) \geq f(x')$, for all trial points x^k, $k \geq 1$;*
3. *if there exists another limit point $x'' \neq x'$, then $f(x'') = f(x')$;*
4. *if the function $f(x)$ has a finite number of local minima in $[a, b]$, then the point x' is locally optimal;*
5. *(Sufficient conditions for convergence to a global minimizer). Let x^* be a global minimizer of $f(x)$ and an iteration number k^* exist such that for all $k > k^*$ the inequality*

$$l_{j(k)} > L_{j(k)} \tag{2.23}$$

holds, where $L_{j(k)}$ is the Lipschitz constant for the interval $[x_{j(k)-1}, x_{j(k)}]$ containing x^* and $l_{j(k)}$ is its estimate. Then, the set of limit points of the sequence $\{x^k\}$ coincides with the set of global minimizers of the function $f(x)$.

Proof Since all the methods mentioned in the Theorem belong to the divide-the-best class of global optimization algorithms introduced in [277], the proofs of assertions 1–5 can be easily obtained as particular cases of the respective proofs in [277, 290].

$\square$

Corollary 2.1 *Assertions 1–5 hold for methods belonging to the GS and using the "Additive" Local Tuning if the condition $l_{i(k)} > r H_{i(k)}$ is fulfilled for all i and k.*

Proof Fulfillment of the condition $l_{i(k)} > r H_{i(k)}$ ensures that: (i) each new trial point x^{k+1} belongs to the interval (x_{t-1}, x_t) chosen for partitioning; (ii) the distances $x^{k+1} - x_{t-1}$ and $x_t - x^{k+1}$ are finite. The fulfillment of these two conditions implies that the methods belong to the class of divide-the-best global optimization algorithms and, therefore, proofs of assertions 1–5 can be easily obtained as particular cases of the respective proofs in [277, 290].

$\square$

Note that in practice, since both ε and δ assume finite positive values, methods using the optimistic local improvement can miss the global optimum and stop in the δ-neighborhood of a local minimizer (see Step 4 of the GS).

Next Theorem ensures existence of the values of the reliability parameter r satisfying condition (2.23), providing so the fact that all global minimizers of $f(x)$ will be determined by the proposed methods without using the a priori known Lipschitz constant.

Theorem 2.2 *For any function $f(x)$ satisfying the Lipschitz condition (2.3) with $0 < L < \infty$ and for methods belonging to the GS and using one of the estimates (2.9), (2.11), (2.16), (2.18) there exists a value r^* such that, for all $r > r^*$, condition (2.23) holds.*

Proof It follows from, (2.9), (2.11), (2.16), (2.18), and the finiteness of $\xi > 0$ that approximations of the Lipschitz constant l_i in the methods belonging to the GS are always positive. Since L in (2.3) is finite and any positive value of the parameter r can be chosen in (2.9), (2.11), (2.16), (2.18), it follows that there exists an r^* such that condition (2.23) will be satisfied for all global minimizers for $r > r^*$.

$\square$

2.5 Numerical Illustrations

Seven series of numerical experiments were executed in [301] (see also [299]) on the following three sets of test functions to compare the described 22 global optimization methods:

1. the widely used set of 20 test functions from [140];
2. 100 randomly generated Pintér's functions from [243];
3. four functions originated from practical problems: first two problems are from [290, page 113] and the other two functions from [123] (see also [120, 296]).

Table 2.3 Number of trials performed by the considered geometric methods without the local improvement on 20 tests from [140]

#	Geom-AL	Geom-GL	Geom-LTM	Geom-LTA	Geom-LTMA
1	595	446	50	44	**35**
2	457	373	49	52	**39**
3	577	522	176	202	**84**
4	1177	1235	57	73	**47**
5	383	444	57	65	**43**
6	301	299	70	73	**50**
7	575	402	53	51	**41**
8	485	481	164	183	**82**
9	469	358	55	57	**41**
10	571	481	55	58	**42**
11	1099	1192	100	109	**78**
12	993	1029	93	96	**68**
13	2833	2174	93	88	**68**
14	379	303	56	60	**39**
15	2513	1651	89	118	**72**
16	2855	2442	102	120	**83**
17	2109	1437	125	171	**122**
18	849	749	55	58	**41**
19	499	377	49	47	**39**
20	1017	166	53	58	**40**
Avg	1036.80	828.05	80.05	89.15	**57.70**

Geometric and information methods with and without the local improvement techniques (optimistic and pessimistic) were tested in these experimental series. In all the experiments, the accuracy of the global search was chosen as $\varepsilon = 10^{-5}(b - a)$, where $[a, b]$ is the search interval. The accuracy of the local search was set as $\delta = \varepsilon$ in the algorithms with the local improvement. Results of numerical experiments are reported in Tables 2.3, 2.4, 2.5, 2.6, 2.7, 2.8, 2.9 and 2.10, where the number of function trials executed until the satisfaction of the stopping rule is presented for each considered method (the best results for the methods within the same class are shown in bold).

The first series of numerical experiments was carried out with geometric and information algorithms without the local improvement on 20 test functions from [140]. Parameters of the geometric methods Geom-AL, Geom-GL, Geom-LTM, Geom-LTA, and Geom-LTMA were chosen as follows. For the method Geom-AL, the estimates of the Lipschitz constants were computed as the maximum between the values calculated as relative differences on 10^{-7}-grid and the values given in [140]. For the methods Geom-GL, Geom-LTM, and Geom-LTMA, the reliability parameter

Table 2.4 Number of trials performed by the considered information methods without the local improvement on 20 tests from [140]

#	Inf-AL	Inf-GL	Inf-LTM	Inf-LTA	Inf-LTMA
1	422	501	46	35	**32**
2	323	373	47	38	**36**
3	390	504	173	72	**56**
4	833	1076	51	56	**47**
5	269	334	59	47	**37**
6	208	239	65	46	**45**
7	403	318	49	38	**37**
8	157	477	163	113	**63**
9	329	339	54	48	**42**
10	406	435	51	42	**38**
11	773	1153	95	78	**75**
12	706	918	88	71	**64**
13	2012	1351	54	54	**51**
14	264	349	55	44	**38**
15	1778	1893	81	82	**71**
16	2023	1592	71	67	**64**
17	1489	1484	128	121	**105**
18	601	684	52	**43**	43
19	352	336	44	34	**33**
20	681	171	55	**39**	39
Avg	720.95	726.35	74.05	58.40	**50.80**

Table 2.5 Results of numerical experiments with the considered geometric and information methods without the local improvement on 100 Pintér's test functions from [243]

Method	Average	StDev	Method	Average	StDev
Geom-AL	1080.24	91.17	Inf-AL	750.03	66.23
Geom-GL	502.17	148.25	Inf-GL	423.19	109.26
Geom-LTM	58.96	9.92	Inf-LTM	52.13	5.61
Geom-LTA	70.48	17.15	Inf-LTA	**36.47**	6.58
Geom-LTMA	**42.34**	6.63	Inf-LTMA	38.10	5.96

$r = 1.1$ was used as recommended in [290]. The technical parameter $\xi = 10^{-8}$ was used for all the methods with the local tuning (Geom-LTM, Geom-LTA, and Geom-LTMA). For the method Geom-LTA, the parameter r was increased with the step equal to 0.1, starting from $r = 1.1$ until all 20 test problems were solved (i.e., for all the problems the algorithm stopped in the ε-neighborhood of a global minimizer).

Table 2.6 Number of trials performed by the considered geometric and information methods without the local improvement on four applied test functions

Method	Test problem				Average
	1	2	3	4	
Geom-AL	37	395	261	332	256.25
Geom-GL	39	388	216	307	237.50
Geom-LTM	37	54	59	232	95.50
Geom-LTA	74	58	68	204	101.00
Geom-LTMA	**33**	**39**	**48**	**137**	**64.25**
Inf-AL	**12**	278	180	187	164.25
Inf-GL	35	333	215	229	203.00
Inf-LTM	25	53	56	212	86.50
Inf-LTA	19	**35**	**40**	165	64.75
Inf-LTMA	24	**35**	**40**	**122**	**55.25**

Table 2.7 Number of trials performed by the considered geometric and information methods with the *optimistic* local improvement on 20 tests from [140]

#	Geom LTIMO	Geom LTIAO	Geom LTIMAO	Inf LTIMO	Inf LTIAO	Inf LTIMAO
1	45	41	**35**	47	**35**	37
2	47	49	**35**	45	**37**	41
3	49	45	**39**	55	**45**	51
4	47	53	**43**	**49**	53	53
5	55	49	**47**	51	**47**	**47**
6	51	49	**45**	47	**43**	47
7	45	45	**39**	49	**37**	39
8	37	41	**35**	**41**	45	47
9	49	51	**41**	51	51	**40**
10	47	49	**41**	51	**43**	43
11	49	53	**45**	55	59	**55**
12	43	53	**35**	53	67	**45**
13	**51**	53	57	**41**	51	55
14	45	45	**43**	49	**43**	45
15	**45**	57	47	**45**	55	53
16	**49**	55	53	**47**	49	53
17	93	**53**	95	59	55	**53**
18	45	47	**37**	49	**41**	44
19	45	43	**35**	46	**33**	35
20	43	45	**37**	49	**35**	39
Avg	49.00	48.80	**44.20**	48.95	46.20	**46.10**

Table 2.8 Number of trials performed by the considered geometric and information methods with the *pessimistic* local improvement on 20 tests from [140]

#	Geom LTIMP	Geom LTIAP	Geom LTIMAP	Inf LTIMP	Inf LTIAP	Inf LTIMAP
1	49	46	**36**	47	38	**35**
2	49	50	**38**	47	37	**35**
3	165	212	**111**	177	**56**	57
4	56	73	**47**	51	56	**46**
5	63	66	**48**	57	47	**38**
6	70	71	**51**	64	46	**45**
7	54	53	**41**	51	39	**38**
8	157	182	**81**	163	116	**99**
9	53	57	**43**	52	52	**43**
10	56	59	**42**	52	43	**39**
11	100	114	**77**	95	78	**72**
12	93	97	**69**	87	73	**64**
13	97	86	**68**	55	52	**50**
14	58	197	**43**	60	46	**42**
15	79	120	**76**	79	82	**70**
16	97	115	**81**	71	66	**60**
17	140	189	**139**	127	129	**100**
18	55	60	**42**	51	**42**	42
19	52	50	**36**	46	33	**32**
20	54	56	**40**	51	**37**	40
Avg	79.85	97.65	**60.45**	74.15	58.40	**52.35**

This situation has happened for $r = 1.8$: the corresponding results are shown in the column Geom-LTA of Table 2.3.

As can be seen from Table 2.3, the performance of the method Geom-LTMA was better than the behavior of the other geometric algorithms tested. The experiments also showed that the additive convolution (Geom-LTA) did not guarantee the proximity of the found solution to the global minimum with the common value $r = 1.1$ used by the other tested methods. With an increased value of the reliability parameter r, the average number of trials performed by this method on 20 tests was also slightly worse than that of the method with the maximum convolution (Geom-LTM), but better than the averages of the methods using global estimates of the Lipschitz constants (Geom-AL and Geom-GL).

Results of numerical experiments with information methods without the local improvement techniques (methods Inf-AL, Inf-GL, Inf-LTM, Inf-LTA, and Inf-LTMA) on the same 20 tests from [140] are shown in Table 2.4. Parameters of the information methods were chosen as follows. The estimates of the Lipschitz constants for the method Inf-AL were the same as for the method Geom-AL. The reliability

Table 2.9 Results of numerical experiments with the considered geometric and information methods with the local improvement techniques on 100 Pintér's test functions from [243]

Optimistic strategy				Pessimistic strategy			
Method	r	Average	StDev	Method	r	Average	StDev
Geom-LTIMO	1.3	49.52	4.28	Geom-LTIMP	1.1	66.44	21.63
Geom-LTIAO	1.9	48.32	5.02	Geom-LTIAP	1.8	93.92	197.61
Geom-LTIMAO	1.4	**45.76**	5.83	Geom-LTIMAP	1.1	**48.24**	14.12
Inf-LTIMO	2.0	48.31	4.29	Inf-LTIMP	2.0	53.06	7.54
Inf-LTIAO	2.1	**36.90**	5.91	Inf-LTIAP	2.0	**37.21**	7.25
Inf-LTIMAO	2.0	38.24	6.36	Inf-LTIMAP	2.0	39.06	6.84

Table 2.10 Number of trials performed by the considered geometric and information methods with the local improvement techniques on four applied test functions

	Method	r	Test problem				Average
			1	2	3	4	
Optimistic LI	GeomLTIMO	6.5	59	55	63	79	64.00
	Geom-LTIAO	1.8	55	**49**	**49**	**41**	**48.50**
	GeomLTIMAO	6.9	63	59	71	75	67.00
	Inf-LTIMO	6.5	49	55	67	77	62.00
	Inf-LTIAO	9.4	**47**	55	71	71	61.00
	Inf-LTIMAO	8.0	55	55	71	73	63.50
Pessimistic LI	Geom-LTIMP	1.1	39	64	71	228	100.50
	Geom-LTIAP	1.8	243	102	63	1254	415.50
	Geom-LTIMAP	1.1	31	46	51	**106**	**58.50**
	Inf-LTIMP	2.0	25	52	58	185	80.00
	Inf-LTIAP	2.3	**18**	36	49	174	69.25
	Inf-LTIMAP	2.0	24	**35**	**43**	134	59.00

parameter $r = 2$ was used in the methods Inf-GL, Inf-LTM, and Inf-LTMA, as recommended in [290, 315, 323]. For all the information methods with the local tuning techniques (Inf-LTM, Inf-LTA, and Inf-LTMA), the value $\xi = 10^{-8}$ was used. For the method Inf-LTA, the parameter r was increased (starting from $r = 2$) up to the value $r = 2.3$ when all 20 test problems were solved.

As can be seen from Table 2.4, the performance of the method Inf-LTMA was better (as also verified for its geometric counterpart) with respect to the other information algorithms tested. The experiments have also shown that the average number of trials performed by the Inf-LTA method with $r = 2.3$ on 20 tests was better than that of the method with the maximum convolution (Inf-LTM).

The second series of experiments (see Table 2.5) was executed on the class of 100 Pintér's test functions from [243] with all geometric and information algorithms without the local improvement (i.e., all the methods used in the first series of experiments).

Parameters of the methods Geom-AL, Geom-GL, Geom-LTM, Geom-LTMA, and Inf-AL, Inf-GL, Inf-LTM and Inf-LTMA were the same as in the first series ($r = 1.1$ for all the geometric methods and $r = 2$ for the information methods). The reliability parameter for the method Geom-LTA was increased again from $r = 1.1$ to $r = 1.8$ (when all 100 problems have been solved). All the information methods were able to solve all 100 test problems with $r = 2$ (see Table 2.5). The average performance of the Geom-LTMA and the Inf-LTA methods was the best among the other considered geometric and information algorithms, respectively.

The third series of the experiments (see Table 2.6) was carried out on four applied test problems from [123, 290]. All the methods without the local improvement used in the previous two series of experiments were tested and all the parameters for these methods were the same as above, except the reliability parameters of the methods Geom-LTA and Inf-LTA. Particularly, the applied problem 4 was not solved by the Geom-LTA method with $r = 1.1$. With the increased value $r = 1.8$, the obtained results (reported in Table 2.6) of this geometric method were worse than the results of the other geometric methods with the local tuning (Geom-LTM and Geom-LTMA). The method Inf-LTA has solved all the four applied problems also with a higher value $r = 2.3$ and was outrun by the Inf-LTMA method.

In the following series of experiments, the local improvement techniques were compared on the same three sets of test functions. In the fourth series (see Table 2.7), six methods (geometric and information) with the optimistic local improvement (methods Geom-LTIMO, Geom-LTIAO, Geom-LTIMAO and Inf-LTIMO, Inf-LTIAO and Inf-LTIMAO) were compared on the class of 20 test functions from [140]. The reliability parameter $r = 1.1$ was used for the methods Geom-LTIMO and Geom-LTIMAO and $r = 2$ was used for the method Inf-LTIMO. For the method Geom-LTIAO r was increased to 1.6, and for the methods Inf-LTIMAO and Inf-LTIAO r was increased to 2.3. As can be seen from Table 2.7, the best average result was shown by the method Geom-LTIMAO (while the Inf-LTIMAO was the best in average among the considered information methods).

In the fifth series of experiments, six methods (geometric and information) using the pessimistic local improvement were compared on the same 20 test functions. The obtained results are presented in Table 2.8. The usual values $r = 1.1$ and $r = 2$ were used for the geometric (Geom-LTIMP and Geom-LTIMAP) and the information (Inf-LTIMP and Inf-LTIMAP) methods, respectively. The values of the reliability parameter ensuring the solution to all the test problems in the case of methods Geom-LTIAP and Inf-LTIAP were set as $r = 1.8$ and $r = 2.3$, respectively. As can be seen from Table 2.8, the "Maximum" and the "Maximum-Additive" local tuning techniques were more stable, and generally allowed us to find the global solution for all test problems without increasing r. The methods Geom-LTIMAP and Inf-LTIMAP showed the best performance with respect to the other geometric and information techniques, respectively.

In the sixth series of experiments, the local improvement techniques were compared on the class of 100 Pintér's functions. The obtained results are presented in Table 2.9. The values of the reliability parameter r for all the methods were increased, starting from $r = 1.1$ for the geometric methods and $r = 2$ for the information methods, until all 100 problems from the class were solved. It can be seen from Table 2.9, that the best average number of trials for both the optimistic and pessimistic strategies was almost the same (36.90 and 37.21 in the case of information methods and 45.76 and 48.24 in the case of geometric methods, for the optimistic and for the pessimistic strategies, respectively). However, the pessimistic strategy seemed to be more stable since its reliability parameter (providing solutions to all the problems) generally remained smaller than that of the optimistic strategy. In average, the Geom-LTMA and the Inf-LTA methods were the best among the other considered geometric and information algorithms, respectively.

Finally, the last, seventh, series of the experiments (see Table 2.10) was executed on the class of four applied test problems, where in the third column the values of the reliability parameter used to solve all the problems are also indicated. Again, as in the previous experiments, the pessimistic local improvement strategy seemed to be more stable in the case of this test set, since the optimistic strategy required a significant increase of the parameter r to determine global minimizers of these applied problems (although the best average value obtained by the optimistic Geom-LTIAO method was smaller than that of the best pessimistic Geom-LTIMAP method).

Let us summarize the results of this section. Univariate derivative-free global optimization has been considered and several numerical methods belonging to the geometric and information classes of algorithms have been proposed and analyzed. Acceleration techniques (see, e.g., [301]) to speed up the global search have been described. They can be used in both the geometric and information frameworks. All of the considered methods automatically switch from the global optimization to the local one and back, avoiding so the necessity to stop the global phase manually. An original mixture of new and traditional computational steps has allowed the authors in [301] to construct 22 different global optimization algorithms having, however, a similar structure. As was shown, 9 instances of this mixture can lead to well-known global optimization methods, and the remaining 13 methods tested in the present section were first introduced in [301]. All of them have been studied theoretically in the previous section and numerically compared on 124 theoretical and applied benchmark tests. It has been shown that the introduced acceleration techniques allow the global optimization methods to significantly speed up the search with respect to some known algorithms.

Chapter 3
Diagonal Approach and Efficient Diagonal Partitions

> *The first rule of any technology used in a business is that automation applied to an efficient operation will magnify the efficiency. The second is that automation applied to an inefficient operation will magnify the inefficiency.*

> Bill Gates

3.1 General Diagonal Scheme

In the rest of this book, the problem of minimization of a multidimensional multiextremal black-box function that satisfies the Lipschitz condition over a multidimensional hyperinterval $D \subset \mathbb{R}^N$ with an unknown Lipschitz constant $0 < L < \infty$ will be considered:

$$f^* = f(x^*) = \min f(x), \quad x \in D, \tag{3.1}$$

where

$$D = [a, b] = \{x \in \mathbb{R}^N : a(j) \leq x(j) \leq b(j), \ 1 \leq j \leq N\}, \quad a, b \in \mathbb{R}^N, \tag{3.2}$$

the objective function $f(x)$ satisfies the Lipschitz condition

$$|f(x') - f(x'')| \leq L \|x' - x''\|, \quad \forall x', x'' \in D, \quad 0 < L < \infty, \tag{3.3}$$

and $\| \cdot \|$ denotes the Euclidean norm. If the function $f(x)$ is non-differentiable, only its values at $x \in D$ can be used in solving problem (3.1)–(3.3). If $f(x)$ is smooth then the function gradient $\nabla f(x)$ can be also evaluated at trial points. It is hereafter assumed that in both the cases performing a trial (i.e., evaluating either $f(x)$ only or both $f(x)$ and $\nabla f(x)$ at a point $x \in D$) is a time-consuming operation.

© The Author(s) 2017
Y.D. Sergeyev and D.E. Kvasov, *Deterministic Global Optimization*,
SpringerBriefs in Optimization, DOI 10.1007/978-1-4939-7199-2_3

In the introductory Chap. 1, some approaches to solving Lipschitz global optimization problems (3.1)–(3.3) have been introduced briefly. It has been observed
that these problems pose a significant computational challenge for the optimizers,
especially in the case where the problem dimension N from (3.3) is high. Several
ideas proposed to tackle multidimensional problems have been also briefly described
in Chap. 1. Among them, the following two approaches to solving these problems
have been discussed. The first one consists of the partition algorithms which subdivide the feasible set D from (3.2) into subsets $D_i \subseteq D$ and approximate (on the
basis of the trial results obtained) behavior of the objective function $f(x)$ over D by
estimating $f(x)$ over each subset D_i. Convergence analysis of many of the partition
algorithms can be performed in a unified manner by using the general divide-the-
best algorithms scheme. The second approach aims to reduce the multidimensional
problem (3.1)–(3.3) to the one-dimensional case. It allows one to apply efficient one-
dimensional algorithms (as, e.g., those described in the previous Chap. 2) to solving
multidimensional problems.

In this chapter, the *diagonal approach* proposed by Pintér in [239–242], unifying,
in a sense, ideas of the two mentioned approaches to solving (3.1)–(3.3), will be
considered in detail. Recall that the diagonal approach can be informally described
as follows. A diagonal algorithm sequentially subdivides the search hyperinterval D
from (3.2) into a set of adaptively generated hyperintervals $D_i = [a_i, b_i]$ with the
vertices a_i, b_i and the main diagonals $[a_i, b_i]$ (i. e., the line segment connecting the
vertices a_i and b_i of the hyperinterval D_i), such that

$$D = \bigcup_{i=1}^{M(k)+\Delta M(k)-1} D_i, \quad D_i \cap D_j = \delta(D_i) \cap \delta(D_j), \quad i \neq j.x \qquad (3.4)$$

In (3.4), $\delta(D_i)$ denotes the boundary of D_i, the value $M = M(k)$ is the number of
hyperintervals D_i at the beginning of the iteration k of the algorithm, and $\Delta M(k) > 1$
is the number of new hyperintervals produced during the subdivision of a hyperinterval at the current kth iteration (the subdivided hyperinterval of the current partition
is substituted by $\Delta M(k)$ new hyperintervals). Formula (3.4) represents the current
diagonal partition $\{D^k\}$ of the search domain D obtained at the kth iteration of the
diagonal algorithm.

In order to decrease the computational efforts needed to describe behavior of the
objective function $f(x)$ at every D_i, $f(x)$ (and $\nabla f(x)$ if available) is evaluated only
at the vertices a_i and b_i of D_i. At each step the 'merit' of each hyperinterval so far
generated is estimated. The higher 'merit' (measured by a real-valued function R_i
called *characteristic*; see Sect. 1.3) of a hyperinterval D_i corresponds to the higher
possibility that the global minimizer x^* of $f(x)$ belongs to D_i. In order to calculate
the characteristic R_i of a multidimensional hyperinterval D_i, some one-dimensional
characteristics (see Chap. 2) can be used as prototypes. After an appropriate transformation they can be applied to the one-dimensional segment being the main diagonal
$[a_i, b_i]$ of the hyperinterval D_i. A hyperinterval having the 'best' characteristic (e.g.,
the largest one) is partitioned by means of a partition operator (diagonal partition

strategy) P, and new trials are performed at two vertices corresponding to the main diagonal of each generated hyperinterval. The concrete choice of the function R_i and the partition strategy P determines each particular diagonal method.

Unless otherwise specified, hereafter in this chapter we shall suppose that the derivatives of the objective function are not available and only one hyperinterval is subdivided at each iteration k of the diagonal algorithm. The latter does not hold, for example, in the DIRECT algorithm mentioned in the previous chapters, where several hyperintervals can be subdivided at each iteration.

In order to give a formal description of the multidimensional diagonal algorithms, the following notations are introduced:

$k \geq 1$—the current iteration number of the diagonal algorithm;

$p(k)$—the total number of function trials executed during the previous $k - 1$ iterations of the algorithm;

$\Delta p(k)$—the number of new function trials performed during the kth iteration of the algorithm;

$\{x^{p(k)}\}$—the sequence of trial points generated by the algorithm in k iterations;

$\{z^{p(k)}\} = \{f(x^{p(k)})\}$—the corresponding sequence of the objective function values;

$M = M(k)$—the total number of hyperintervals of the current partition of the search domain at the beginning of the kth iteration;

$\Delta M(k)$—the number of new hyperintervals produced during the kth iteration of the algorithm;

$\{D^k\} = \{D_i^k\} = \{D_i\}, 1 \leq i \leq M$,—the current partition of the initial hyperinterval D from (3.2) into hyperintervals $D_i = [a_i, b_i]$ from (3.4). For each hyperinterval D_i, the coordinates of the vertices a_i and b_i of its main diagonal $[a_i, b_i]$ and the corresponding values of $f(a_i)$ and $f(b_i)$ are evaluated.

The general scheme (*diagonal scheme*) of the diagonal algorithms (see [239–242, 279]) can be described as follows:

Step 1. *Initialization:* Perform the first two trials at the vertices of the initial domain D, i.e., $x^1 = a$, $z^1 = f(x^1)$ and $x^2 = b$, $z^2 = f(x^2)$, and set the current number of the function trials $p(1) := 2$. Let the current partition of D be $\{D^1\} := \{D\}$, and set the current number of the generated hyperintervals $M(1) := 1$. Set the iteration counter $k := 1$.

Suppose now that $k \geq 1$ iterations of the algorithm have already been executed. The iteration $k + 1$ consists of the following steps.

Step 2. *Characteristics Calculations:* For each hyperinterval $D_i = [a_i, b_i], 1 \leq i \leq M(k)$, from the current partition $\{D^k\}$, calculate its characteristic (cf. (2.19), (2.19))

$$R_i = R(D_i^k, \{x^{p(k)}\}, \{z^{p(k)}\}, v), \quad 1 \leq i \leq M(k), \tag{3.5}$$

where v is a vector of parameters of the algorithm.

Step 3. *Hyperinterval Selection:* Select a hyperinterval $D_t = [a_t, b_t]$ from $\{D^k\}$ having an index t, $1 \leq i \leq M(k)$, such that

$$t = \arg \max_{1 \leq i \leq M(k)} R_i. \tag{3.6}$$

Step 4. *Stopping Criterion:* If

$$\|a_t - b_t\| \leq \varepsilon \|a - b\|, \tag{3.7}$$

where $\varepsilon > 0$ is a prescribed accuracy, a and b are the vertices of the main diagonal of the search hyperinterval D, t is from (3.6), and $\| \cdot \|$ denotes the Euclidean norm, then **Stop**. Take as an estimate of the global minimum of $f(x)$ over D the value

$$z_k^* = \min\{z : z \in \{z^{p(k)}\}\}$$

attained at the point

$$x_k^* = \arg \min\{f(x^j) : x^j \in \{x^{p(k)}\}\}.$$

Otherwise, go to **Step 5**.

Step 5. *New Partition:* Subdivide the hyperinterval D_t into $\Delta M(k)$, $\Delta M(k) > 1$, new hyperintervals by means of a partition strategy P, i.e.,

$$\{D^{k+1}\} = P(\{D^k\}, \{x^{p(k)}\}, \{z^{p(k)}\}, \nu),$$

in such a way that the new partition $\{D^{k+1}\}$ is constructed substituting the hyperinterval D_t by the newly generated hyperintervals as follows:

$$D_{i(k)} = D_{i(k+1)}, \quad i(k) \neq t, \quad i(k+1) \neq t,$$

$$D_{t(k)} = D_{t(k+1)} \cup \left(\bigcup_{i=M(k)+1}^{M(k)+\Delta M(k)-1} D_{i(k+1)} \right).$$

Step 6. *New Trials:* Perform $\Delta p(k)$ new trials at the vertices corresponding to the main diagonal of the new generated hyperintervals. Set $p(k+1) := p(k) + \Delta p(k)$, $M(k+1) := M(k) + \Delta M(k) - 1$. Increase the iteration counter $k := k + 1$ and go to **Step 2**.

Let us first give some comments on the diagonal scheme described above and then consider some possible ways for both the calculation of the characteristic R_i from (3.5) and the subdivision of hyperintervals having the maximal characteristics. Notice that usually the partition strategy P subdividing a hyperinterval D_t selected in Step 5 of the diagonal scheme works with the so-called *point selection function* [240]

$$S_t = S(D_t^k, \{x^{p(k)}\}, \{z^{p(k)}\}, v), \quad 1 \leq t \leq M(k). \tag{3.8}$$

This function determines a point within the hyperinterval D_t (normally, on the main diagonal of D_t), and the subdivision of D_t is then performed by a number of hyperplanes passed through the chosen point and parallel to the boundary hyperplanes of D_t. As observed, e.g., in [240, 242], the point selection function S_t plays a technical role (although this role is important for an efficient implementation of diagonal algorithms) and there always exist possibilities to implement this function in order to guarantee the contraction of the hyperintervals selected for partitioning (as required by the divide-the-best scheme; see, e.g., [277]). Some examples of the point selection function and diagonal partition strategies often used in diagonal algorithms will be considered further.

Notice also that the stopping criterion used in Step 4 of the diagonal scheme can be substituted by other criteria (testing, for instance, the volume of the hyperinterval D_t selected for partitioning or the exhaustion of the available computing resources, such as a maximal number of function trials).

The diagonal algorithms belong to the class of divide-the-best algorithms [277]. Thus, on the one hand, general convergence theory developed for analysis of divide-the-best algorithms in [277] can be successfully applied to analysis of the diagonal algorithms, too. On the other hand, the diagonal approach provides a natural generalization (see, e.g., [240–242]) of many one-dimensional algorithms to the multidimensional case.

In fact, the points a_i and b_i being the vertices of the main diagonal of every hyperinterval D_i of the current partition of D can be considered as the end points of a one-dimensional interval. Therefore, a one-dimensional algorithm can be used to describe behavior of the objective function $f(x)$ over the main diagonal on the basis of the information $f(a_i)$, $f(b_i)$ obtained as a result of the trials at the points a_i, b_i. The obtained information is then generalized from the one-dimensional to the multidimensional region D_i, and a conclusion about behavior of the objective function over the whole multidimensional hyperinterval D_i is made.

Thus, one-dimensional divide-the-best algorithms (see, e.g., [133, 135] and Chap. 2 for their description) including Piyavskij–Shubert's, Strongin's, and local tuning algorithms from Sects. 2.2 and 2.4 can be extended to the multidimensional case. In the formulae for the characteristics of these one-dimensional methods (see, e.g., formulae (1.6), (2.6), (2.19) and (2.20) for the cases of geometric and information methods, respectively; other examples are given in [240–242]), the distance between two extreme points of a one-dimensional interval should be expressed by the Euclidean norm. Points $x \in (a_i, b_i)$ from the interior of a one-dimensional interval $[a_i, b_i]$ (see, e.g., [240–242]) should be analogously generated on the main diagonal $[a_i, b_i]$ of a multidimensional hyperinterval D_i.

For example, Piyavskij–Shubert's method can be generalized to the multidimensional case by means of the diagonal scheme as follows. For a hyperinterval $D_i = [a_i, b_i]$ from (3.4) the characteristic R_i and the point S_i from (3.8) are calculated as (see [175, 240, 242]; cf. (2.6) and (2.7)):

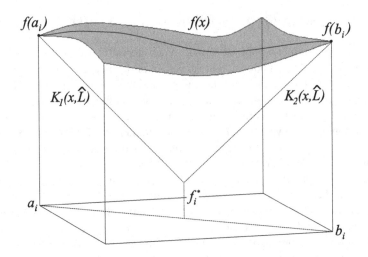

Fig. 3.1 Estimation f_i^* of the lower bound of a Lipschitz function $f(x)$ over a hyperinterval $D_i = [a_i, b_i]$ in diagonal algorithms

$$R_i = -\frac{f(a_i) + f(b_i)}{2} + \hat{L}\frac{\|b_i - a_i\|}{2}, \quad 1 \leq i \leq M(k), \tag{3.9}$$

$$S_i = \frac{a_i + b_i}{2} - \frac{f(b_i) - f(a_i)}{2\hat{L}} \times \frac{b_i - a_i}{\|b_i - a_i\|}, \quad 1 \leq i \leq M(k), \tag{3.10}$$

where $\hat{L}$ is an overestimate of the Lipschitz constant L.

In this case, instead of constructing a lower bounding function for $f(x)$ over the whole search domain D (cf. Fig. 1.4) or over its parts D_i, a minorant function for $f(x)$ is built only over the one-dimensional segment $[a_i, b_i]$ of hyperinterval D_i (see Fig. 3.1). This minorant function is the maximum of two linear functions $K_1(x, \hat{L})$ and $K_2(x, \hat{L})$ passing with the slopes $\pm\hat{L}$ through the vertices a_i and b_i of the main diagonal $[a_i, b_i]$ of D_i. The characteristic value R_i from (3.9) is, therefore, calculated (similarly to (1.6) and (2.6)) as the minimum value f_i^* (multiplied by -1 for a purely technical reason, accordingly to Step 3 of the diagonal scheme) of the one-dimensional minorant function at the intersection of the lines $K_1(x, \hat{L})$ and $K_2(x, \hat{L})$ (see Fig. 3.1).

As has been shown in [242], a valid estimate f_i^* becomes the lower bound of $f(x)$ over the whole hyperinterval D_i if an overestimate $\hat{L}$ of the Lipschitz constant L from (3.3) satisfies the following inequality:

$$\hat{L} \geq 2L. \tag{3.11}$$

Thus, the lower bound of the objective function over the whole hyperinterval $D_i \subseteq D$ can be estimated by considering $f(x)$ only along the main diagonal $[a_i, b_i]$ of D_i. Note also that condition (3.11) is sufficient to prove the global convergence of Piyavskij–Shubert's algorithm extended to the multidimensional case by means of the diagonal scheme using formulae (3.9) and (3.10) (see, e.g., [240–242]). In the following, a more precise inequality than (3.11) will be obtained and the corresponding convergence conditions will be demonstrated.

Another example of the one-dimensional algorithm suitable for the extension to the multidimensional case is Strongin's information algorithm (see Sects. 2.2 and 2.4). Its extension is performed by introducing the following characteristic R_i and the point S_i (see [176, 216, 240–242]):

$$R_i = H^k \|a_i - b_i\| + \frac{(f(a_i) - f(b_i))^2}{H^k \|a_i - b_i\|} - 2(f(a_i) + f(b_i)), \qquad (3.12)$$

$$S_i = \frac{a_i + b_i}{2} - \frac{f(b_i) - f(a_i)}{2H^k} \times \frac{b_i - a_i}{\|b_i - a_i\|}, \qquad (3.13)$$

where H^k is an adaptive estimate of the global Lipschitz constant L from (3.3). At each iteration $k \geq 1$ of the algorithm, it can be calculated, for example, as follows (see [240]; cf. (2.8)):

$$H^k = H(k) = (4 + \frac{C}{k}) \max_{1 \leq i \leq M(k)} \frac{|f(a_i) - f(b_i)|}{\|a_i - b_i\|}, \qquad (3.14)$$

where the constant $C > 0$ is a parameter of the method. As demonstrated in [240–242], if, starting from an iteration number k^* during minimization of a Lipschitz (with the Lipschitz constant L) function $f(x)$, the inequality

$$H(k) \geq 4L, \quad k > k^*, \qquad (3.15)$$

holds for the estimate (3.14) of L, then, the set X' of the limit points of the trial sequence $\{x^{p(k)}\}$ generated by the diagonally extended Strongin's algorithm and the set X^* of the global minimizers of $f(x)$ coincide.

3.2 Analysis of Traditional Diagonal Partition Schemes

Let us now discuss the diagonal approach from the viewpoint of the partition strategy P (see both Step 5 of the diagonal scheme and formula (3.8)). Since each trial of $f(x)$ is assumed to be a time-consuming operation, one wants to obtain a solution of the problem by evaluating $f(x)$ at a minimal number of trial points. The usage of the diagonal approach has the goal to decrease the computational efforts needed to describe behavior of $f(x)$ over every hyperinterval D_i by evaluating the function at only two vertices of D_i instead of evaluating it at all 2^N vertices. Traditionally, two strategies for hyperinterval partitioning are used in diagonal algorithms: 2^N-*Partition* (see, e.g., [107, 109, 150, 154, 175, 205, 216, 240, 242]) and *Bisection* (see, e.g., [117, 150, 175, 242, 278]). In this section, we consider both of them in detail.

The diagonal 2^N-Partition strategy operates as follows. A point S_t from (3.8) is chosen on the main diagonal of the hyperinterval D_t, where the index t is from (3.6). As already mentioned, the choice of the point S_t is determined by each particular realization of a diagonal algorithm. The hyperinterval $D_{t(k)}$ is split into

$$\Delta M(k) = 2^N$$

hyperintervals (where N is the problem dimension and $\Delta M(k)$ is the number of new hyperintervals produced during the kth iteration of the diagonal algorithm)

$$D_{t(k+1)}, \quad D_{M(k)+1}, \quad \ldots, \quad D_{M(k)+\Delta M(k)-1}, \tag{3.16}$$

$$t(k+1) = t(k) = t,$$

by means of N hyperplanes passed through the point S_t and parallel to the boundary hyperplanes of D_t. Thus, the hyperinterval D_t is substituted by 2^N new hyperintervals numbered as indicated in (3.16).

After partitioning, the objective function $f(x)$ is evaluated at the vertices a_j and b_j of all new hyperintervals. Note that new trials are performed at

$$\Delta p(k) = 2^{N+1} - 3$$

points where $\Delta p(k)$ is the number of new trials executed during the kth iteration of the diagonal algorithm. In fact, for each of the 2^N hyperintervals, $f(x)$ is to be evaluated at two vertices, but the vertex S_t is such that

$$S_t = a_{M(k)+\Delta M(k)-1} = b_{t(k+1)}$$

and $f(x)$ has already been evaluated at the vertices $a_{t(k)}$ and $b_{t(k)}$ of the initial hyperinterval $D_{t(k)}$ during previous iterations.

In Fig. 3.2, an example of partitioning a two-dimensional hyperinterval D_t by using the 2^N-Partition strategy is given (D_t is presented before and after the splitting). The hyperinterval D_t is substituted by $\Delta M(k) = 4$ new hyperintervals

$$D_{t(k+1)}, \quad D_{M(k)+1}, \quad D_{M(k)+2}, \quad D_{M(k)+3}.$$

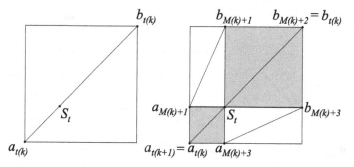

Fig. 3.2 Partition of a two-dimensional hyperinterval D_t executed by the diagonal 2^N-Partition strategy

For instance, the new hyperintervals $D_{t(k+1)}$ and $D_{M(k)+2}$, defined by the vertices $a_{t(k)}$, S_t and S_t, $b_{t(k)}$, respectively, are shown in light gray. The vertices where the objective function $f(x)$ is evaluated are indicated by black dots.

Performing expensive function trials at $2^N - 3$ points during each iteration can impose a too high computational demand on solving the problem (3.1)–(3.3) by a diagonal algorithm using the 2^N-Partition strategy. This limitation can be overcome by the diagonal bisection partition strategy (see [117, 150, 242, 279]). This strategy also determines a point S_t (by means of a point selection function from (3.8)) on the main diagonal of the hyperinterval D_t selected for partitioning (the index t is from (3.6)). But, in contrast to the 2^N-Partition strategy, $f(x)$ is not evaluated at S_t. The subdivision is executed by one hyperplane orthogonal to the longest edge of D_t and passing through the point S_t. Thus, the hyperinterval $D_{t(k)}$ is substituted by $\Delta M(k) = 2$ new hyperintervals $D_{t(k+1)}$, $D_{M(k)+1}$ with vertices:

$$a_{t(k+1)} = a_{t(k)},$$

$$b_{t(k+1)} = (b(1), b(2), \ldots, b(i-1), S_t(i), b(i+1), \ldots, b(N)),$$

$$a_{M(k)+1} = (a(1), a(2), \ldots, a(i-1), S_t(i), a(i+1), \ldots, a(N)),$$

$$b_{M(k)+1} = b_{t(k)},$$

where $a(j)$ and $b(j)$ are the j-th coordinates, $1 \leq j \leq N$, of the vectors $a_{t(k)}$ and $b_{t(k)}$, respectively, and the index i is given by the following formula

$$i = \arg\min \max_{1 \leq j \leq N} |b(j) - a(j)|.$$

Note that we use the term Bisection just to underline the fact of subdivision of the selected hyperinterval into two new hyperintervals. This term does not imply that the newly generated hyperintervals should have the same volume.

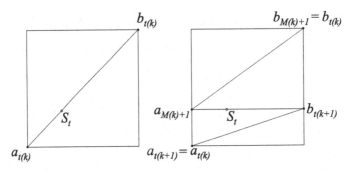

Fig. 3.3 Partition of a two-dimensional hyperinterval D_t executed by the diagonal bisection strategy

After partitioning, the objective function $f(x)$ is evaluated only at

$$\Delta p(k) = 2$$

points corresponding to the vertices $a_{M(k)+1}$ and $b_{t(k+1)}$. Thus, the number of trials to be executed during every iteration does not depend on the problem dimension N. In Fig. 3.3, an illustration of the subdivision of a two-dimensional hyperinterval D_t executed by the diagonal bisection strategy is represented (as before, black dots indicate trial points).

Both the 2^N-Partition and Bisection strategies appeared to be quite efficient from the point of view of the number of trials performed when applied at each particular iteration of a diagonal algorithm. However, as shown in [279], both these strategies generate too many redundant trial points in the course of the algorithm execution (irrespective of the form of the characteristic $R(\cdot)$ that determines which hyperinterval is to be subdivided at each iteration). This redundancy can limit the application of the 2^N-Partition and Bisection strategies when hard black-box functions are to be minimized. Let us consider in more detail (following [279]) the difficulties arising when these traditional diagonal partitions are used.

First of all, in both strategies, their high performance is expected to be ensured by evaluating the function $f(x)$ only at the two vertices corresponding to the main diagonal of each newly generated hyperinterval. Unfortunately, it turns out that each hyperinterval contains more than two trial points in both strategies. For example, during the subdivision of a hyperinterval $D_{t(k)}$ by means of the 2^N-Partition strategy (see Fig. 3.2), the number of function evaluations is the following: (i) the objective function is evaluated at all the 2^N vertices of the two new hyperintervals defined by the vertices $a_{t(k)}$, S_t and S_t, $b_{t(k)}$; (ii) for all the other new hyperintervals, $f(x)$ is evaluated at least at $2^{N-1} + 1$ vertices (see [279, Theorem 2.1]).

The bisection strategy does not maintain the property of evaluating $f(x)$ at only two vertices of each hyperinterval either. For example, it can be seen from Fig. 3.3 that the new hyperinterval defined by the vertices $a_{t(k+1)}$, $b_{t(k+1)}$ has the objective function evaluated at three points $a_{t(k+1)}$, $b_{t(k+1)}$, and $a_{M(k)+1}$. Analogously, the new

Fig. 3.4 A partition of the search hyperinterval D after four iterations executed by a diagonal algorithm using the 2^N-Partition strategy

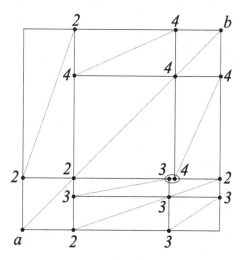

hyperinterval defined by the vertices $a_{M(k)+1}, b_{M(k)+1}$ has $f(x)$ evaluated at three points $a_{M(k)+1}, b_{M(k)+1}$, and $b_{t(k+1)}$. The following general result holds: during the subdivision of a hyperinterval $D_{t(k)}$ by the bisection strategy, the two new hyperintervals have at least three vertices where the objective function $f(x)$ is evaluated (see [279, Theorem 2.2]).

Thus, during every iteration, both the strategies, 2^N-Partition and Bisection, produce hyperintervals where the objective function is evaluated in more than two vertices. In their turn, these hyperintervals can be subsequently subdivided, and the number of redundant evaluations will increase.

The second problem with the 2^N-Partition and Bisection strategies is the loss of information about the proximity of the vertices of hyperintervals generated at different iterations, which leads to an unnecessary evaluation of $f(x)$ at close points (see, e.g., [175, 279, 290, 293]). In the worst case, the vertices of different hyperintervals coincide, and, thus, the objective function is evaluated twice at the same point. In Fig. 3.4, an illustration of this problem for the 2^N-Partition strategy is presented. Each digit is the number of the iteration at which $f(x)$ has been evaluated at the corresponding point (at the first iteration, $f(x)$ has been evaluated at the vertices a and b). It can be seen that two trial points generated during the third and the fourth iterations are very close to one another (the corresponding points are circled in Fig. 3.4). In many cases, it would be sufficient to evaluate $f(x)$ only at one of these points. In spite of this fact, 2^N-Partition strategy performs this redundant work (namely, evaluates $f(x)$ and stores the information obtained twice) and then loses the information about the points vicinity, since these points have been generated during different iterations. Moreover, the distance between these points can be less than the accuracy ε in the stopping rule of a diagonal algorithm (see (3.7)), and the algorithm is not able to recognize this situation efficiently.

A similar effect can be verified if the bisection strategy is employed. Figure 3.5 shows a set of hyperintervals D_i obtained after performing seven partitions of the

Fig. 3.5 A partition of the search hyperinterval D after eight iterations executed by a diagonal algorithm using the bisection strategy

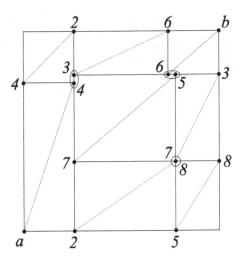

initial hyperinterval D. The close points where the third and fourth and the fifth and sixth iterations have been performed are circled. Note that the two points corresponding to the seventh and eighth iterations coincide: a diagonal method using the bisection strategy would evaluate $f(x)$ at the same point twice and store the results of the trials twice in different areas of the computer memory. Moreover, if for the 2^N-Partition strategy the points S_t (see formula (3.8) and Fig. 3.2) cannot be generated twice, in the bisection strategy all the trial points inside D (i. e., the points that do not belong to the border of D) can be generated and stored twice.

Thus, both the 2^N-Partition and Bisection strategies, appearing to be efficient when considered only at each separated iteration, can produce a high number of redundant trial points, independently of the form of the characteristic $R(\cdot)$ determining a hyperinterval to be subdivided. The main reasons for the appearance of redundant trial points are the same for both strategies: (i) during each iteration, the objective function is evaluated at more than two vertices of every new hyperinterval D_j, among which only two trial points are used to calculate the characteristic R_j of the hyperinterval under consideration; (ii) the strategies are not able to establish links between hyperintervals generated during different iterations efficiently, and, thus, they lose the information about the proximity of trial points in the multidimensional space. The worst case is the coincidence of trial points when $f(x)$ is evaluated twice at the same point and the results of the trials are stored twice in different areas of the computer memory.

Therefore, the meshes of trial points generated by the 2^N-Partition and Bisection strategies, traditionally used in diagonal algorithms, do not respond in a satisfactory manner to the requirement of evaluating the objective function at a minimal number of points in D. The redundant calculations of $f(x)$ lead to both the slowing down the partitioning procedure and the excessive growth of information to be stored in the computer memory.

3.3   Non-redundant Diagonal Partition Strategy

In this section, an efficient partition strategy originally proposed in [279] (see also [274, 290]) which overcomes the limitations of traditional diagonal partition strategies is described. This strategy produces partitions with hyperintervals having exactly two vertices where $f(x)$ is evaluated. Moreover, as shown in [170, 279], a special indexation of the hyperintervals can be proposed in order to establish links between hyperintervals having common facets but generated during different iterations. The usage of this strategy allows one to avoid the duplication of trial points and, therefore, it is called *efficient* or *non-redundant* diagonal partition strategy.

We start our description with two examples illustrating partitions of a two- and a three-dimensional search hyperintervals by means of the non-redundant partition strategy. Let us first consider a two-dimensional example (see Fig. 3.6) by using the terms 'interval' and 'sub-interval' to denote two-dimensional rectangular domains. In this example, ten initial partitions of the admissible interval $D = [a, b]$ from (3.2) produced by a diagonal algorithm using the efficient strategy are given. Trial points of $f(x)$ are represented by black dots. The numbers around these dots indicate iterations at which the objective function is evaluated at the corresponding points. Sub-intervals to be subdivided at each iteration are shown in light gray. Recall that at each iteration only one hyperinterval is supposed to be subdivided.

In Fig. 3.6a, the situation after the first two iterations is presented. At the first iteration, the objective function $f(x)$ is evaluated at the vertices a and b of the search interval $D = [a, b]$. At the next iteration, the interval D is subdivided into equal three sub-intervals. This subdivision is performed by two lines (hyperplanes) orthogonal to the longest edge of D and passing through points u and v (see Fig. 3.6a). The objective function is evaluated at both the points u and v.

Suppose that the interval shown in light gray in Fig. 3.6a is chosen for the further partitioning. Thus, at the third iteration, three smaller sub-intervals appear (see Fig. 3.6b). It seems that a trial point of the third iteration is redundant for the interval (shown in light gray in Fig. 3.6b) selected for the next splitting. But in reality, Fig. 3.6c demonstrates that one of the two points of the fourth iteration (the iteration number around it is enclosed in brackets) coincides with the point 3 at which $f(x)$ has already been evaluated. Therefore, there is no need to evaluate $f(x)$ at this point again, since the function value obtained at the previous iteration can be reused. This value can be stored in a special database and is simply retrieved when necessary without re-evaluation of the function (we shall return to this question later). Figure 3.6d illustrates the situation after 11 iterations. Among 22 points at which the objective function is supposed to be evaluated, there are 5 repeated points (the iteration numbers around these points are enclosed in brackets). That is, $f(x)$ is evaluated 17 rather than 22 times. Note also that the number of generated intervals (equal to 21) is greater than the number of trial points (equal to 17). Such a difference will become more pronounced in the course of further subdivisions, and the number of trial points with reused (not re-evaluated!) function values will increase.

(a)   **(b)**

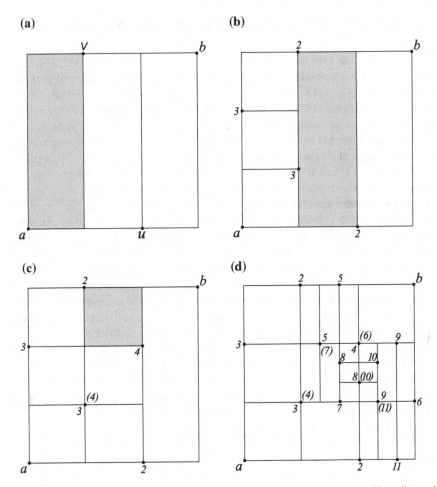

Fig. 3.6 An example of subdivisions of a two-dimensional search interval D executed by a diagonal algorithm using the non-redundant partition strategy (sub-intervals to be subdivided at each iteration of the algorithm are shown in textitlight gray, the numbers in brackets indicate the iterations at which the corresponding function values are not evaluated and the results obtained during previous iterations are used)

Let us now consider a three-dimensional example (see Fig. 3.7), paying a particular attention to the mutual positions of the trial points generated during different iterations (recall that the 2^N-Partition and Bisection strategies suffer from the appearance of redundant trials). As before, trial points of $f(x)$ are represented by black dots and are marked with the number of iterations during which these points have been generated, hyperintervals to be subdivided are shown in light gray. In Fig. 3.7a, the situation after the first two iterations is presented. The trials of the first iteration are executed at the vertices a and b of the search hyperinterval $D = [a, b]$. Then, the first subdivision is performed during the second iteration and $f(x)$ is evaluated at

(a)   **(b)**

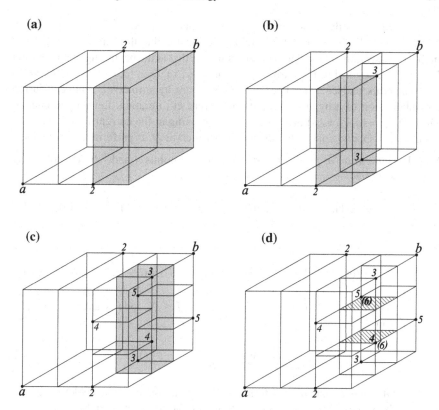

Fig. 3.7 An example of subdivisions of a three-dimensional search hyperinterval D executed by a diagonal algorithm using the non-redundant partition strategy (hyperintervals to be subdivided are shown in light gray)

the points indicated by the number 2 (see Fig. 3.7a). The situation obtained after the third iteration is shown in Fig. 3.7b. After the fourth and fifth iterations, we have the partition presented in Fig. 3.7c. The trial points of these iterations are indicated by the numbers 4 and 5, respectively. During the sixth iteration, the light gray hyperinterval in Fig. 3.7c is subdivided into three sub-hyperintervals (by two hyperplanes drawn by hatching in Fig. 3.7d). It can be seen that all the vertices determining the three new sub-hyperintervals (the iteration numbers around these points are enclosed in brackets) have been generated during previous iterations (see Fig. 3.7d from top to bottom): they are 3 and 5, then 5 and 4, and then 4 and 3. The points where $f(x)$ is to be evaluated at the sixth iteration coincide with the points generated during the fourth and fifth iterations (see the points in the hyperplanes drawn by hatching in Fig. 3.7d). The function $f(x)$ is not re-evaluated at these points since the corresponding values can be taken from data obtained at the fourth and fifth iterations. Thus, the sixth partition has been obtained *for free*.

Now we are ready to present the general scheme of the efficient diagonal partition strategy. Without loss of generality, we assume that the admissible region D in (3.2) is an N-dimensional hypercube. Suppose that at the beginning of an iteration $k \geq 1$ of the algorithm the current partition $\{D^k\}$ of $D = [a, b]$ consists of $M(k)$ hyperintervals, and let $\Delta M(k)$ be the number of new hyperintervals produced during the subdivision of a hyperinterval at the current kth iteration. Let a hyperinterval $D_t = D_{t(k)} = [a_{t(k)}, b_{t(k)}]$ be chosen for partitioning at the current iteration k. The operation of subdivision of the selected hyperinterval D_t is performed as follows.

Step 1. *Generation of points u and v:* Determine points u and v by the following formulae

$$u = (a(1), \ldots, a(i-1), a(i) + \frac{2}{3}(b(i) - a(i)), a(i+1), \ldots, a(N)), \quad (3.17)$$

$$v = (b(1), \ldots, b(i-1), b(i) + \frac{2}{3}(a(i) - b(i)), b(i+1), \ldots, b(N)), \quad (3.18)$$

where $a(j) = a_t(j)$, $b(j) = b_t(j)$, $1 \leq j \leq N$, and i is given by the equation

$$i = \arg\min \max_{1 \leq j \leq N} |b(j) - a(j)|. \quad (3.19)$$

Obtain (evaluate or read from a special database, described below in detail) the values of the objective function $f(x)$ at the points u and v.

Step 2. *Hyperinterval Subdivision:* Subdivide the hyperinterval D_t into

$$\Delta M = 3$$

new equal hyperintervals by two parallel hyperplanes that are perpendicular to the longest edge i of D_t and pass through the points u and v (as represented in Fig. 3.8 where the hyperinterval D_t is shown before and after its subdivision).

Step 3. *New Partition:* Construct the new partition $\{D^{k+1}\}$ by substituting the hyperinterval D_t with the newly generated hyperintervals as follows (see Fig. 3.8 where the superscripts indicate the current iteration number and the subscripts indicate the indices of hyperintervals of the current partition of D):

$$D_{i(k)} = D_{i(k+1)}, \quad i(k) \neq t, \quad i(k+1) \neq t,$$

$$D_{t(k)} = D_{t(k+1)} \cup (\bigcup_{i=M(k)+1}^{M(k)+\Delta M(k)-1} D_{i(k+1)}),$$

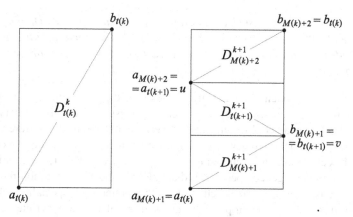

Fig. 3.8 The non-redundant diagonal partition strategy subdivides a hyperinterval D_t into three equal hyperintervals and trials are performed at exactly two vertices for each of these hyperintervals

where the new hyperintervals $D_{t(k+1)}$, $D_{M(k)+1}$, and $D_{M(k)+2}$ are determined by the vertices of their main diagonals (see Fig. 3.8)

$$a_{t(k+1)} = a_{M(k)+2} = u, \quad b_{t(k+1)} = b_{M(k)+1} = v, \tag{3.20}$$

$$a_{M(k)+1} = a_{t(k)}, \quad b_{M(k)+1} = v, \tag{3.21}$$

$$a_{M(k)+2} = u, \quad b_{M(k)+2} = b_{t(k)}. \tag{3.22}$$

Set $M(k+1) := M(k) + \Delta M(k) - 1$ and increase the iteration counter $k := k + 1$ (recall that only one hyperinterval is supposed to be subdivided at each iteration).

Note that due to (3.20)–(3.22) and in contrast to the 2^N-Partition and Bisection strategies, the condition

$$a_i(j) < b_i(j), \quad j = 1, 2, \ldots, N, \tag{3.23}$$

is not satisfied for all hyperintervals $D_i \subset D$, and the orientations of the main diagonals of D_i in $\mathbb{R}^N$ may be different. However, their orientations are not arbitrary. It has been shown in [279] that the vertices of the generated hyperintervals are located in such a way that the disadvantages of the 2^N-Partition and Bisection strategies resulting in the generation of redundant trial points can be eliminated. The following two theorems demonstrated in [279] establish a theoretical base of the non-redundant diagonal partition strategy.

Theorem 3.1 *Every hyperinterval D_t having $f(x)$ evaluated only at the vertices a_t and b_t is split by using the non-redundant diagonal partition strategy into three hyperintervals with exactly two vertices where the function $f(x)$ is evaluated.*

Theorem 3.2 *There exists such an indexation of the hyperintervals D_i, obtained in the course of subdivisions by using the non-redundant diagonal partition strategy, that the index of a hyperinterval D_i can be used to calculate the coordinates of its vertices a_i and b_i where the function $f(x)$ is evaluated.*

The described efficient diagonal strategy resolves the problems found in the 2^N-Partition and Bisection strategies (see Sect. 3.2). First, at every hyperinterval, the function $f(x)$ is evaluated exactly at two vertices (Theorem 3.1). Moreover, by evaluating $f(x)$ at only two new points, three hyperintervals are created, in contrast with the bisection strategy, generating two hyperintervals. This fact can influence positively the convergence speed of a diagonal method using the non-redundant strategy because of a faster constriction of the hyperintervals (especially in the case when the diagonal method stops if the main diagonal of the hyperinterval to be partitioned becomes smaller than an accuracy ε; see formula (3.7)).

Second, the indexation mentioned in Theorem 3.2 allows one to establish links between the hyperintervals obtained during the current partitioning and the hyperintervals having joint facets with them but generated during previous iterations. One very important consequence of this result is the complete absence of duplication of the trial points present in the traditional strategies discussed in Sect. 3.2. Let us consider this important issue in more detail (see [169, 176, 279, 288]).

Every vertex where $f(x)$ is evaluated can belong to different (up to 2^N) hyperintervals (see Figs. 3.6d and 3.7d). The indexation of the hyperintervals avoids the need to store the coordinates a_i and b_i for a hyperinterval D_i in the part of memory regarding D_i, since it is possible to calculate these coordinates by knowing the index of D_i (see [279]). Thus, the coordinates of the vertices and the related description information (including the corresponding function value) can be stored in a separate database (hereafter, called *vertex database*). In this case, the hyperintervals can have only pointers to the vertices and do not duplicate the coordinates a_i and b_i and the related description information.

When a hyperinterval D_t is subdivided, the indices of the three new sub-intervals are known. Therefore, the coordinates of the corresponding vertices (3.20)–(3.22) can be found easily. Instead of an immediate evaluation of the values $f(u)$ and $f(v)$ at the new trial points u and v, first, the existence of these points in the database is verified, because $f(u)$ and/or $f(v)$ could be evaluated during previous iterations. Thus, three cases are possible:

(i) both the points u and v exist in the vertex database;
(ii) only one of them exists;
(iii) both of them are absent.

In the first case, we only retrieve the data (namely, the function values) from the database. In the second case, we read one value and create a new record in the database for the absent point (say, u), evaluate the corresponding function value $f(u)$, and record it in the element created. In the third case, these operations are executed for both the points u and v.

Thus, the objective function value at a vertex is calculated only once, stored in the database, and read when required. It is important to note that a trial point generated

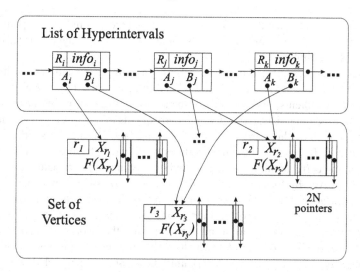

Fig. 3.9 A data structure for implementation of the vertex database used by the non-redundant diagonal partition strategy

by the non-redundant partition strategy can belong to several (up to 2^N) hyperintervals. Since the time-consuming operation of the function evaluation is replaced by a significantly faster operation of reading (up to 2^N times) the function values from the vertex database, the introduced partition strategy considerably speeds up the search and also leads to saving computer memory. In this context, it is particularly significant that the advantage of a diagonal method using this efficient strategy (in terms of the number of function trials) will increase with the problem dimension N (cf. Figs. 3.6 and 3.7)

An implementation of the vertex database is not a simple task since both the operations of retrieving elements from the database and insertion of new elements into it must be realized efficiently. Therefore, a special data structure should be adopted in order to implement these operations. An illustrative example of such a data structure is represented in Fig. 3.9. In this structure, the data describing hyperintervals and vertices are separated into two parts—in a *list of hyperintervals* and in a *set of vertices*, respectively. The list of hyperintervals is a linear list (see, e.g., [55] for a discussion on linear lists) each element of which corresponds to a hyperinterval D_i of the current partition of D and contains information regarding the hyperinterval D_i as, e.g., its characteristic R_i, the number of subdivisions of each side of D performed to obtained the hyperinterval under consideration, and other description information designated by $info_i$ in Fig. 3.9. The hyperinterval D_i of the list of hyperintervals has only pointers A_i and B_i to the elements of the set of vertices, containing the coordinates of its vertices a_i and b_i (together with the related information about the objective function), respectively. This allows us to avoid the duplication of information regarding the vertices of hyperintervals. For example, if two hyperintervals D_i and D_k have the common vertices $b_i = b_k$, then the related information is memo-

rized in a unique record of the set of vertices (see the record r_3 in Fig. 3.9, pointed by
the pointers B_i and B_k from the elements i and k of the list of hyperintervals) and,
therefore, is not duplicated.

The set of vertices consists of the records r_j containing information regarding the
coordinates X_{r_j} of the vertices and the related information $F(X_{r_j})$ (e.g., function
values and/or gradients). The records of the set of vertices form (by means of $2N$
pointers) bidirectional lists, each of which corresponds to one of the N coordinates.
The management of these lists uses the following fact. During the subdivision of
a hyperinterval $D_t = [a_t, b_t]$ the new vertices u and v (see formulae (3.17)–(3.18))
differ from the corresponding vertices a_t and b_t only in one coordinate (namely, the
coordinate i given by formula (3.19)). Thus, in order to establish whether a record
with the point u (or the point v) exists in the vertex database, only a few elements
corresponding to the points with all the coordinates but the coordinate i equal to
the coordinates of a_t (or b_t) are to be examined. In other words, it is sufficient to
confront the coordinate $u(i)$ (or $v(i)$) with the i-th coordinates of the points belonging
to the line $x(i) = a_t(i)$ (or $x(i) = b_t(i)$), until these values are bounded by $b_t(i)$ (or
$a_t(i)$). That is why the elements of the set of vertices are organized in N different
lists, i. e., each list for a coordinate, which are updated at every insertion of a new
element and are maintained ordered, in the order of growth of the corresponding
coordinates. Since condition (3.23) is not necessarily satisfied for all the coordinates
of a hyperinterval of the current partition, each of the N lists is bidirectional that
provides an efficient searching/inserting procedure for any possible orientation of a
hyperinterval in the space.

The mentioned data structure is at the heart of an implementation of the spe-
cialized vertex database, discussed in detail in [168, 288]. Note that the database can be used not only for solving problem (3.1)–(3.3), but also for solving
a general problem of the minimal description of the structure of a vector func-
tion $\mathscr{F}(x) = (f_1(x), f_2(x), \ldots, f_q(x))$ over an N-dimensional hyperinterval, if this
description is provided by applying diagonal algorithms based on the non-redundant
partition strategy. Here, the term 'minimal description' means that it is required to
obtain some knowledge about $\mathscr{F}(x)$ by evaluating it at a minimal number of trial
points $x \in D$ (see [279] for details).

In this context, we should recall (see Fig. 1.7 in the introductory Chap. 1) that the
efficient diagonal strategy can be also viewed as a procedure generating a series of
curves similar to traditional space-filling curves (see, e.g., [238, 263, 315, 323])—
adaptive diagonal curves (see [176, 279, 288]). They join the vertices a and b of
the initial hyperinterval D and remain continuous during the whole process of partitioning. Each of these curves is constructed on the main diagonals of hyperintervals
obtained during the current partition of D. Particularly, the initial adaptive diagonal
curve coincides with the main diagonal $[a, b]$ of the hyperinterval D. Then, every
partitioning of a hyperinterval D_t substitutes the segment of the curve, corresponding
to the main diagonal $[a_t, b_t]$ of D_t, with the polygonal line connecting the points a_t,
v, u, and b_t (see the right picture in Fig. 3.8). Thus, a new curve is generated, which
differs from the previous one only within the subdomain D_t. This reflects the idea
of adaptive local partition of the curve within the selected subdomain (without mod-

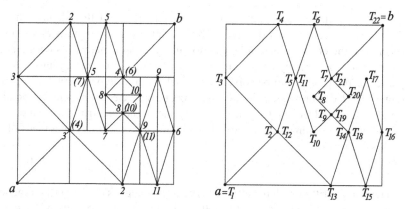

Fig. 3.10 A partition of a two-dimensional interval D after ten subdivisions executed by the non-redundant diagonal partition strategy and the corresponding adaptive diagonal curve constructed on the main diagonals of sub-intervals of this partition

Fig. 3.11 Adaptive diagonal curves become denser in the vicinity of the global minimizers of $f(x)$ if the selection of hyperintervals for partitioning is realized appropriately

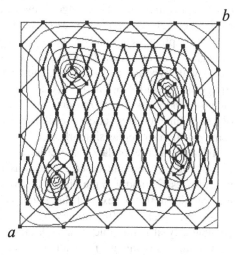

ifying other segments) at the current iteration. In Fig. 3.10, a partition (and obtained diagonals) of a two-dimensional interval D after ten subdivisions executed by the non-redundant diagonal partition strategy (left picture) and the corresponding adaptive diagonal curve (right picture) are presented. Trial points of $f(x)$ are indicated by black dots. In the left picture (corresponding to the partition from Fig. 3.6), the numbers around the dots indicate iterations at which the objective function is evaluated at the corresponding points (if enclosed in brackets, the corresponding function value has been read from the vertex database). The curve starts at the point $a = T_1$, goes through the points $T_2, T_3, \ldots, T_{21}$, and finishes at the point $b = T_{22}$.

Usually in numerical algorithms some regular approximations of the space-filling curves with an a priori given level of the subdivision depth (the same over the whole region) are used (see, e.g., [263, 315, 317, 323] and approximations to Peano curves

considered in Sect. 1.3). In contrast, the order of partition of the adaptive diagonal curves is different within different sub-hyperintervals of D and is determined
by specifying the characteristic $R(\cdot)$. A particular diagonal method using the non-
redundant partition strategy constructs its own series of curves, taking into account
properties of the problem to be solved. If the selection of hyperintervals for partitioning is realized appropriately in the algorithm, the curve condenses in the vicinity
of the global minimizers of $f(x)$. In Fig. 3.11, one of the adaptive diagonal curves
generated during minimization of the following function of two variables (see [142])

$$f(x) = (x_1^2 + x_2 - 11)^2 + (x_1 + x_2^2 - 7)^2, \quad x = (x_1, x_2) \in D = [-6, 6]^2,$$

by a diagonal algorithm with the efficient partition strategy is depicted (the function
level curves are also drawn in Fig. 3.11). It can be seen that the density of the curve
in Fig. 3.11 increases in the neighborhood of the four global minimizers.

To conclude, we can notice (see [280]) that the non-redundant strategy developed
for the diagonal methods can be successfully applied for the so-called *one-point-
based algorithms*, too. These algorithms (see, e.g., [60, 109, 148, 154, 205, 235,
236]) adaptively subdivide the search domain (3.2) in smaller subregions (hyperintervals, simplicies, etc.) and evaluate $f(x)$ from (3.1) at only one point within each
subregion. For example, the DIRECT optimization method discussed in Chap. 1
works with hyperintervals and uses the center-sampling one-point-based partition
strategy in its work (see Fig. 1.5 in Chap. 1). As shown in [280], if a one-point-based
method subdivides the search domain (3.2) into hyperintervals and evaluates the
objective function at a vertex of each hyperinterval, redundant trials can appear for
some partition strategies similarly to traditional diagonal methods (see Sect. 3.2).

To avoid this redundancy, the efficient diagonal partition strategy can be used
in the case of one-point-based methods as follows. Instead of evaluating $f(x)$ at
two vertices—u and v in the diagonal scheme (3.17)–(3.18)—it is proposed to do
this initially at the vertex a of the region D and then at the vertices u from (3.17)
during every splitting (the point v from (3.18) is used just for partitioning goals).
Analogously, it is possible to start the function evaluations from the vertex b and
then continue to evaluate $f(x)$ at the vertices v from (3.18) while using the vertices
u from (3.17) for partitioning only. To verify whether the function $f(x)$ has been
already evaluated at each trial point, the fast procedure developed previously for the
efficient diagonal partition strategy can be successfully used also in this case.

To illustrate this *non-redundant* one-point-based partition strategy, let us consider an example presented in Fig. 3.12 assuming that a single iteration consists of
the subdivision of only one hyperinterval. Black dots represent the trial points and
the numbers around these dots indicate iterations at which these trial points have
been generated. The terms 'interval' and 'sub-interval' are again used here to denote
two-dimensional rectangular domains. In Fig. 3.12a, the situation after the first two
iterations is presented. Particularly, at the second iteration, the interval D is partitioned into three sub-intervals of equal area (equal volume in a general case). This
subdivision is performed by two lines (hyperplanes) orthogonal to the longest edge
of D (see Fig. 3.12a). The trial is performed only at the point denoted by number

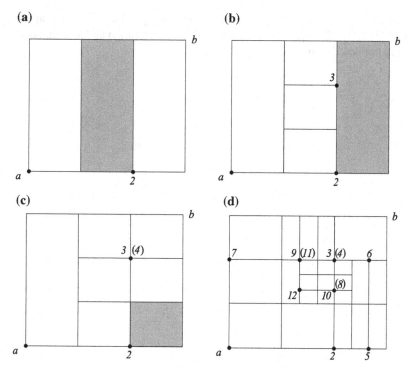

Fig. 3.12 An example of subdivisions by the non-redundant one-point-based partitioning strategy

2. Supposing that the interval shown in light gray in Fig. 3.12a is chosen for the further partitioning, at the third iteration three smaller sub-intervals are generated (see Fig. 3.12b). As one can see from Fig. 3.12c, the trial point of the fourth iteration coincides with the point 3 at which the trial has already been executed. Therefore, there is no need to perform a new (costly) evaluation of $f(x)$ at this point, since the values obtained at the previous iteration and stored in the vertex database can be used. Figure 3.12d illustrates the situation after 12 iterations. It can be seen from this figure that 23 intervals have been generated by 9 trial points only.

Thus, in this section, an efficient adaptive diagonal partition strategy generating regular trial meshes and overcoming difficulties of traditional strategies has been described and its properties examined. It is important to mention that the advantages of this strategy in diagonal algorithms (and in suitably constructed one-point-based methods) become more pronounced when the problem dimension increases. The diagonal partition scheme considered in this section opens interesting perspectives for creating new fast global optimization algorithms. First, popular one-dimensional methods (including those discussed in Chap. 2) may be efficiently extended to the multidimensional case by using this scheme. As an illustration, two multidimensional diagonal algorithms for solving Lipschitz global optimization problems based on the non-redundant strategy will be presented in the following chapter. Second, the described partition strategy may be successfully parallelized by the approach from [135, 323], allowing one to obtain a further acceleration and to estimate theoretically the possible speed-up.

Chapter 4
Global Optimization Algorithms Based on the Non-redundant Partitions

No matter how good you get, you can always get better and that's the exciting part.

Tiger Woods

4.1 Multiple Estimates of the Lipschitz Constant

In this chapter, we present two Lipschitz global optimization algorithms using both (diagonal and one-point-based) versions of the non-redundant partition strategy. The first method (described in this section) is diagonal and derivative-free, whereas the second one (considered in the next section) is a one-point-based method that uses the gradients $\nabla f(x)$ of the objective function $f(x)$ in its work. The respective Lipschitz constants for $f(x)$ and $\nabla f(x)$ are estimated in the algorithms by using a set of possible values varying from zero to infinity.

Remember that the idea of the multiple estimation of the Lipschitz constant originates from the DIRECT algorithm [154], where a center-sampling hyperinterval partition strategy is used. Due to its relative simplicity and a satisfactory performance on several test functions and applied problems, DIRECT has been widely adopted in practical applications (see, e.g., [12, 20, 21, 30, 44, 57, 106, 141, 215, 262, 338]).

As has been pointed out by several authors (see, e.g., [57, 93, 141, 153, 233]), some aspects of DIRECT can limit its area of application. First of all, it is difficult to apply for DIRECT some meaningful stopping criterion, such as, for example, stopping on achieving a desired accuracy of the solution. This happens because DIRECT does not use a single estimate of the Lipschitz constant but works with a set of possible values of L. Although several attempts to introduce a reasonable criterion of arrest have been made (see, e.g., [20, 57, 106, 141]), termination of the search process caused by exhaustion of the available computing resources (such as

© The Author(s) 2017
Y.D. Sergeyev and D.E. Kvasov, *Deterministic Global Optimization*,
SpringerBriefs in Optimization, DOI 10.1007/978-1-4939-7199-2_4

the maximal number of function evaluations) remains the most appropriate one for practical engineering applications.

Another important observation regarding DIRECT is related to the partition and sampling strategies adopted by the algorithm (see [154]) because its simplicity turns into some problems. As has been outlined, e.g., in [57, 93, 233], DIRECT is quick to locate regions of some local optima but slow to converge to the global one. This can happen for several reasons. The first one is a redundant (especially in high dimensions, see [153]) partition of hyperintervals along all longest sides. Partitioning along only one long side suggested in [153] can accelerate convergence in high dimensions in certain cases but it can remain still slow for highly multiextremal problems.

The next cause of DIRECT's slow convergence can be an excessive partition of many small hyperintervals located in the vicinity of local minimizers which are not global ones. Some problems associated with the DIRECT parameter ε preventing the method from subdividing too small hyperintervals are also known. For example, it has been shown in [93], that the original DIRECT algorithm is sensitive to the additive scaling of the objective function (that is, DIRECT is neither homogeneous, see [79] nor strongly homogeneous, see [352]). In order to eliminate this excessive sensitivity, different scaling techniques for the algorithm's parameter have been proposed (see, e.g., [93, 141, 192]). Unfortunately, in this case the algorithm becomes too sensitive to tuning such a parameter, especially for hard black-box global optimization problems (3.1)–(3.3).

In [12, 338], another modification to DIRECT, called 'aggressive DIRECT', has been proposed. It subdivides all hyperintervals with the smallest function value for each hyperinterval size. This results in more hyperintervals partitioned at every iteration and the number of hyperintervals to be subdivided grows significantly. In [106, 107], the opposite idea which is more biased toward local improvement of the objective function has been studied. Results obtained in [106, 107] demonstrate that this modification seems to be more suitable for low-dimensional problems with a single global minimizer and a few local minimizers.

The possibility to use DIRECT together with local optimization methods have been also studied (see, e.g., [153, 194, 195], but this requires the usage of a separate local optimizer and the problem of a correct switching between the global and local phases of the search arises. The balancing of the global and local information within the same global scheme has been studied in [193, 233, 289].

Finally, it should be mentioned that DIRECT—like all center-sampling partitioning schemes—uses relatively poor information about the behavior of the objective function $f(x)$ over each hyperinterval. This information is obtained by evaluating $f(x)$ only at one central point of each hyperinterval without considering the adjacent hyperintervals. Due to this fact, DIRECT can manifest a slow convergence (as has been highlighted in [151]) in cases where the global minimizer lies at the boundary of the admissible region D from (3.2). Other types of partitions, as the simplicial one, can be used in the DIRECT-type methods (see, e.g., [233–235]. This type of partitioning is more informative with respect to the center-sampling hyperinterval partitioning, but is more sophisticated for the implementation. It has also a relatively long and complex preparation phase (related to the initial subdivision of the search region in simplicies) before the global search can start. The diagonal approach can be a good compromise in this sense, since a diagonal algorithm uses more information about the objective function over a hyperinterval than a central-sampling method and is computationally more light than a simplicial algorithm.

4.2 Derivative-Free Diagonal Method MULTL

The goal of this section is to present a global optimization algorithm (the diagonal algorithm using a set of Lipschitz constants MULTL from [289], which would be oriented (in contrast with the algorithm from [106, 107]) on solving hard multidimensional multiextremal black-box problems (3.1)–(3.3). The MULTL method uses a smart technique balancing local and global information for selection of hyperintervals to be subdivided. It is unified with the efficient diagonal partition strategy described in the previous chapter. Then, a novel procedure for an estimation of lower bounds of the objective function over hyperintervals is combined with the idea (introduced in DIRECT) of the usage of a set of Lipschitz constants instead of a unique estimate. As demonstrated by the broad numerical results in [289], application of this algorithm to minimizing hard multidimensional black-box functions leads to a significant improvement with respect to the DIRECT algorithm [154] and its modification [106, 107].

This section is organized as follows. First, a theoretical background of MULTL—a technique for lower bounding the objective function over hyperintervals and a procedure for selection of 'non-dominated' hyperintervals for eventual partitioning—is presented. Then, the algorithm is introduced and its convergence analysis is performed. Results of numerical experiments executed on more than 800 test functions are presented and discussed in the numerical Sect. 4.4.

4.2.1 Theoretical Background of MULTL: Lower Bounds

Let us suppose that at some iteration $k > 1$ of the algorithm the admissible region D from (3.2) has been partitioned into hyperintervals $D_i \in \{D^k\}$ defined by their main diagonals $[a_i, b_i]$ (see formula (3.4)). Due to the general diagonal scheme from Sect. 3.1, at least one of these hyperintervals should be selected for further partitioning. Recall that in order to make this selection, the algorithm estimates the goodness (or, in other words, characteristics) of the generated hyperintervals with respect to the search for a global minimizer. The best (in some predefined sense) characteristic obtained over some hyperinterval D_t corresponds to a higher possibility to find the global minimizer within D_t. This hyperinterval is subdivided at the next iteration of the algorithm. Naturally, more than one 'promising' hyperinterval can be partitioned at each iteration.

One of the possible characteristics of a hyperinterval can be an estimate of the lower bound of $f(x)$ over this hyperinterval. Once all lower bounds for all hyperintervals of the current partition $\{D^k\}$ have been calculated, the hyperinterval with the smallest lower bound can be selected for the further partitioning.

We remind that different approaches to finding lower bounds of $f(x)$ have been proposed in the literature (see, e.g., [109, 148, 154, 205, 207, 242, 246, 306, 323] and Chaps. 1 and 2 of the present book) for solving Lipschitz global optimization problems. For example, given an overestimate $\hat{L}$ of the Lipschitz constant L, a lower bounding (or minorant) function for $f(x)$ can be constructed as the upper envelope of a set of N-dimensional circular cones with the slope $\hat{L}$ (as in Fig. 1.4 of Chap. 1; see also, e.g., [140, 207, 246]). Trial points of $f(x)$ are chosen as the vertices of the cones. At each iteration, the global minimizer of the minorant function is determined and chosen as a new trial point. As has been discussed in Chap. 1, finding such a point requires analyzing the intersections of all the cones and, generally, is a difficult and time-consuming task, especially in high dimensions.

Since in (3.4) we consider a partition of D into hyperintervals, each cone can be constructed over the corresponding hyperinterval, independently of the other cones. This allows one (see, e.g., [107, 140, 154, 175, 242]) to avoid the necessity of establishing the intersections of the cones and to simplify the lower bound estimation. For example, the multidimensional DIRECT algorithm [154] uses one cone with symmetry axis passed through a central point of a hyperinterval for lower bounding $f(x)$ over this hyperinterval. The lower bound is obtained at one of the vertices of the hyperinterval. This approach is simple, but it gives a too rough estimate of the minimal function value over the hyperinterval.

As was already discussed, a more accurate estimate is achieved when two trial points over a hyperinterval are used for constructing a minorant function for $f(x)$, as is done in the diagonal approach. The objective function is evaluated at two vertices of a hyperinterval $D_i = [a_i, b_i]$. Instead of constructing a minorant function for $f(x)$ over the whole hyperinterval D_i, we use a minorant function for $f(x)$ only over the one-dimensional segment $[a_i, b_i]$. Given an overestimate $\hat{L}$ of the Lipschitz constant L, this minorant function is the maximum of two linear functions $K_1(x, \hat{L})$ and $K_2(x, \hat{L})$ passing with the slopes $\pm\hat{L}$ through the vertices a_i and b_i (see Fig. 3.1 in the previous chapter). The lower bound of $f(x)$ over the diagonal $[a_i, b_i]$ of D_i is calculated at the intersection of the lines $K_1(x, \hat{L})$ and $K_2(x, \hat{L})$ and is given by the following formula (see [175, 240, 242]; cf. (3.9)):

$$R_i = R_i(\hat{L}) = 0.5(f(a_i) + f(b_i) - \hat{L}\|b_i - a_i\|), \qquad 0 < L \le \hat{L} < \infty. \qquad (4.1)$$

A valid estimate of the lower bound of $f(x)$ over D_i can be obtained from (4.1) if an appropriate estimate $\hat{L}$ of the Lipschitz constant L is used. In [242], it has been shown that the estimate

$$\hat{L} \ge 2L \qquad (4.2)$$

guarantees that the value R_i from (4.1) is the lower bound of $f(x)$ for the whole hyperinterval D_i. Thus, the lower bound of the objective function over the whole hyperinterval $D_i \subseteq D$ can be estimated by considering $f(x)$ only along the main diagonal $[a_i, b_i]$ of D_i. In the following Theorem 4.1 from [289], a more precise than (4.2) condition is obtained ensuring that

$$R_i(\hat{L}) \leq f(x), \quad x \in D_i.$$

Theorem 4.1 *Let L be the known Lipschitz constant for $f(x)$ from (3.3), $D_i = [a_i, b_i]$ be a hyperinterval of the current partition $\{D^k\}$ from (3.4), and f_i^* be the minimum function value over D_i, i.e.,*

$$f_i^* = f(x_i^*), \quad x_i^* = \arg \min_{x \in D_i} f(x). \tag{4.3}$$

If an overestimate $\hat{L}$ in (4.1) satisfies the inequality

$$\hat{L} \geq \sqrt{2}L, \tag{4.4}$$

then $R_i(\hat{L})$ from (4.1) is the lower bound of $f(x)$ over D_i, i.e., $R_i(\hat{L}) \leq f_i^*$.

Proof Since x_i^* from (4.3) belongs to D_i and $f(x)$ satisfies the Lipschitz condition (3.3) over D_i, then the following inequalities hold

$$f(a_i) - f_i^* \leq L\|a_i - x_i^*\|,$$

$$f(b_i) - f_i^* \leq L\|b_i - x_i^*\|.$$

By summing up these inequalities and using the fact that

$$\max_{x \in D_i}(\|a_i - x\| + \|b_i - x\|) \leq \sqrt{2}\|b_i - a_i\|,$$

we obtain

$$f(a_i) + f(b_i) \leq 2f_i^* + L(\|a_i - x_i^*\| + \|b_i - x_i^*\|) \leq$$

$$\leq 2f_i^* + L \max_{x \in D_i}(\|a_i - x\| + \|b_i - x\|) \leq 2f_i^* + \sqrt{2}L\|b_i - a_i\|.$$

Then, from the last inequality and (4.4) we can deduce that the following estimate holds for the value R_i from (4.1)

$$R_i(\hat{L}) \leq \frac{1}{2}(2f_i^* + \sqrt{2}L\|b_i - a_i\| - \hat{L}\|b_i - a_i\|) =$$

$$= f_i^* + \frac{1}{2}\underbrace{(\sqrt{2}L - \hat{L})}_{\leq 0}\|b_i - a_i\| \leq f_i^*.$$

The theorem has been proved. □

Theorem 4.1 allows us to obtain a more precise lower bound R_i with respect to [240–242] where estimate (4.2) is considered to construct diagonally extended geometric algorithms.

4.2.2   Theoretical Background of MULTL: Finding Non-dominated Hyperintervals

Let us now consider a diagonal partition $\{D^k\}$ of the admissible region D, generated by the efficient diagonal partition strategy. Let a positive value $\tilde{L}$ be chosen as an estimate of the Lipschitz constant L from (3.3) and lower bounds $R_i(\tilde{L})$ of the objective function over hyperintervals $D_i \in \{D^k\}$ be calculated by formula (4.1). Using the obtained lower bounds of $f(x)$, the relation of domination can be established between every two hyperintervals of the current partition $\{D^k\}$ of the admissible region D.

Definition 4.1 Given an estimate $\tilde{L} > 0$ of the Lipschitz constant L from (3.3), a hyperinterval $D_i \in \{D^k\}$ *dominates* a hyperinterval $D_j \in \{D^k\}$ with respect to $\tilde{L}$ if

$$R_i(\tilde{L}) < R_j(\tilde{L}).$$

Definition 4.2 Given an estimate $\tilde{L} > 0$ of the Lipschitz constant L from (3.3), a hyperinterval $D_t \in \{D^k\}$ is said to be *non-dominated with respect to* $\tilde{L} > 0$ if for the chosen value $\tilde{L}$ there is no other hyperinterval in $\{D^k\}$ which dominates D_t.

Each hyperinterval $D_i = [a_i, b_i] \in \{D^k\}$ can be represented (see [289]) by a dot in a two-dimensional diagram (see Fig. 4.1) similar to that used in DIRECT for representing hyperintervals with $f(x)$ evaluated only at one central point (see [154]). However, since we work with the diagonal methods, the diagram should be constructed using different (with respect to DIRECT) values on both coordinate axes. The horizontal coordinate d_i and the vertical coordinate F_i of a dot are defined as follows

$$d_i = \frac{\|b_i - a_i\|}{2}, \quad F_i = \frac{f(a_i) + f(b_i)}{2}, \quad a_i \neq b_i. \tag{4.5}$$

It should be mentioned that each point (d_i, F_i) in the diagram can correspond to several hyperintervals with the same length of the main diagonals and the same sum of the function values at their vertices.

For the sake of illustration, let us consider a hyperinterval D_A with the main diagonal $[a_A, b_A]$ represented by the dot A in Fig. 4.1. If we assume that an estimate of the Lipschitz constant is equal to $\tilde{L}$ (such that condition (4.4) is satisfied), a lower bound of $f(x)$ over the hyperinterval D_A is given by the value $R_A(\tilde{L})$ from (4.1). This value is the vertical coordinate of the intersection point of the line passed through the point A with the slope $\tilde{L}$ and the vertical coordinate axis (see Fig. 4.1). In fact, as can be seen from (4.1), intersection of the line with the slope $\tilde{L}$ passed through any dot representing a hyperinterval in the diagram of Fig. 4.1 and the vertical coordinate axis gives us the lower bound (4.1) of $f(x)$ over the corresponding hyperinterval.

Notice that the points on the vertical axis ($d_i = 0$) do not represent hyperintervals. The axis is used to express such values as lower bounds, the current estimate of the minimal function value, etc. It should be highlighted that the current estimate f_{min} of

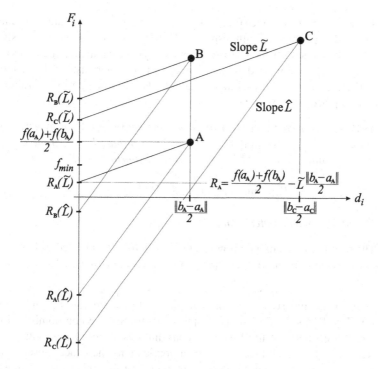

Fig. 4.1 A two-dimensional diagram representing hyperintervals of the current partition of D generated by the non-redundant diagonal strategy and the lower bounds of $f(x)$ over them corresponding to two different estimates $\tilde{L}$ and $\hat{L}$ of the Lipschitz constant L

the global minimum value is always smaller than or equal to the vertical coordinate of the lowest dot (dot A in Fig. 4.1). Note also that due to (4.5) vertex at which this value has been obtained can belong to a hyperinterval different from that represented by the lowest dot in the diagram.

By using this graphical representation, it is easy to determine whether a hyperinterval dominates (with respect to a given estimate of the Lipschitz constant) some other hyperinterval from a partition $\{D^k\}$. For example, for the estimate $\tilde{L}$ the following inequalities are satisfied (see Fig. 4.1)

$$R_A(\tilde{L}) < R_C(\tilde{L}) < R_B(\tilde{L}).$$

Therefore, with respect to $\tilde{L}$ the hyperinterval D_A (dot A in Fig. 4.1) dominates both hyperintervals D_B (dot B) and D_C (dot C), while D_C dominates D_B. If our partition $\{D^k\}$ consists only of these three hyperintervals, the hyperinterval D_A is non-dominated with respect to $\tilde{L}$.

If a higher estimate $\hat{L} > \tilde{L}$ of the Lipschitz constant is considered (see Fig. 4.1), the hyperinterval D_A still dominates the hyperinterval D_B with respect to $\hat{L}$, since $R_A(\hat{L}) < R_B(\hat{L})$. But D_A in turn is dominated by the hyperinterval D_C with respect to

$\hat{L}$, because $R_A(\hat{L}) > R_C(\hat{L})$ (see Fig. 4.1). Thus, for the chosen estimate $\hat{L}$ the unique non-dominated hyperinterval with respect to $\hat{L}$ is D_C, and not D_A as previously.

As can be happened for the estimate $\tilde{L}$, some hyperintervals (as the hyperinterval D_B in Fig. 4.1) are always dominated by other hyperintervals, independently of the chosen estimate of the Lipschitz constant L. The following result formalizing this fact takes place.

Lemma 4.1 *Given a partition* $\{D^k\}$ *of* D *and the subset* $\{D^k\}_d$ *of hyperintervals having the main diagonals equal to* $d > 0$, *for any estimate* $\tilde{L} > 0$ *of the Lipschitz constant a hyperinterval* $D_t \in \{D^k\}_d$ *dominates a hyperinterval* $D_j \in \{D^k\}_d$ *if and only if*

$$F_t = \min\{F_i : D_i \in \{D^k\}_d\} < F_j, \tag{4.6}$$

where F_i *and* F_j *are calculated using (4.5).*

Proof The result of this lemma follows immediately from (4.1) since all hyperintervals under consideration have the same length of their main diagonals, i.e.,
$$\|b_i - a_i\| = d. \qquad \square$$

There also exist hyperintervals (for example, the hyperintervals D_A and D_C represented in Fig. 4.1 by the dots A and C, respectively) that are non-dominated with respect to one estimate of the Lipschitz constant L and dominated with respect to another estimate of L. Since in practical applications the exact Lipschitz constant (or its valid overestimate) is often unknown, the following idea inspired by DIRECT [154] is adopted.

At each iteration $k > 1$ of MULTL, various estimates of the Lipschitz constant L from zero to infinity are chosen for lower bounding $f(x)$ over hyperintervals. The lower bound of $f(x)$ over a particular hyperinterval is calculated by formula (4.1). Note that since all possible values of the Lipschitz constant are considered, condition (4.4) is automatically satisfied and no additional multipliers are required for an estimate of the Lipschitz constant in (4.1). Examination of the set of possible estimates of the Lipschitz constant leads us to the following definition.

Definition 4.3 A hyperinterval $D_t \in \{D^k\}$ is called *non-dominated* if there exists an estimate $0 < \tilde{L} < \infty$ of the Lipschitz constant L such that D_t is non-dominated with respect to $\tilde{L}$.

In other words, non-dominated hyperintervals are hyperintervals over which $f(x)$ has the smallest lower bound for a particular estimate of the Lipschitz constant. For example, in Fig. 4.1 the hyperintervals D_A and D_C are non-dominated.

Let us now make some observations that will allow us to identify the set of non-dominated hyperintervals. First of all, only hyperintervals D_t satisfying condition (4.6) can be non-dominated. In the two-dimensional diagram (d_i, F_i), where d_i and F_i are from (4.5), such hyperintervals are located at the bottom of each group of points with the same horizontal coordinate, i.e., with the same length of the main diagonals. For example, in Fig. 4.2 these points are designated as A (the largest interval), B, C, E, F, G, and H (the smallest interval).

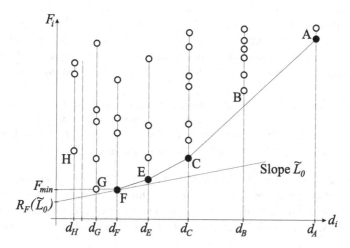

Fig. 4.2 Dominated hyperintervals are represented by white dots and non-dominated hyperintervals are represented by *black* dots

It is important to notice that not all hyperintervals satisfying (4.6) are non-dominated. For example (see Fig. 4.2), the hyperinterval D_H is dominated (with respect to any positive estimate of the Lipschitz constant L) by any of the hyperintervals D_G, D_F, D_E, or D_C. The hyperinterval D_G is dominated by D_F. In fact, as follows from (4.1), among several hyperintervals with the same sum of the function values at their vertices, larger hyperintervals dominate smaller ones with respect to any positive estimate of L. Finally, the hyperinterval D_B is dominated either by the hyperinterval D_A (for example, with respect to $\tilde{L}_1 \geq \tilde{L}_{AC}$, where $\tilde{L}_{AC}$ corresponds to the slope of the line passed through the points A and C in Fig. 4.2), or by the hyperinterval D_C (with respect to $\tilde{L}_2 < \tilde{L}_{AC}$).

Note that if an estimate $\tilde{L}$ of the Lipschitz constant is chosen, it is easy to indicate the hyperinterval with the smallest lower bound of $f(x)$, i.e., the non-dominated hyperinterval with respect to $\tilde{L}$. To do this, it is sufficient to position a line with the slope $\tilde{L}$ below the set of dots in the two-dimensional diagram representing hyperintervals of $\{D^k\}$, and then to shift it upwards. The first dot touched by the line indicates the desirable hyperinterval. For example, in Fig. 4.2 the hyperinterval D_F represented by the point F is a non-dominated hyperinterval with respect to $\tilde{L}_0$, since over this hyperinterval $f(x)$ has the smallest lower bound $R_F(\tilde{L}_0)$ for the given estimate $\tilde{L}_0$ of the Lipschitz constant.

Let us now examine various estimates of the Lipschitz constant L from zero to infinity. When a small (close to zero) positive estimate of L is chosen, an almost horizontal line is considered in the two-dimensional diagram representing hyperintervals of a partition $\{D^k\}$. The dot with the smallest vertical coordinate F_{min} (and the largest horizontal coordinate if there are several such dots) is the first to be touched by this line (the case of the dot F in Fig. 4.2). Therefore, a hyperinterval (or hyperintervals) represented by this dot is non-dominated with respect to the chosen estimate of L

and, consequently, non-dominated in the sense of Definition 4.3. By repeating this procedure with higher estimates of the Lipschitz constant (that is, considering lines with higher slopes) all non-dominated hyperintervals can be identified. In Fig. 4.2 the hyperintervals represented by the dots F, E, C, and A are non-dominated hyperintervals.

This procedure can be formalized in terms of the algorithm known as Jarvis March (or gift wrapping; see, e.g., [250]), being an algorithm for identifying the convex hull of the dots. Thus, the following result identifying the set of non-dominated hyperintervals for a given partition $\{D^k\}$ has been proved.

Theorem 4.2 *Let each hyperinterval* $D_i = [a_i, b_i] \in \{D^k\}$ *be represented by a dot with horizontal coordinate* d_i *and vertical coordinate* F_i *defined in (4.5). Then, hyperintervals that are non-dominated in the sense of Definition 4.3 are located on the lower-right convex hull of the set of dots representing the hyperintervals.*

We conclude our theoretical consideration of MULTL with the following remarks. As has been shown in [279], the lengths of the main diagonals of hyperintervals generated by the non-redundant partition strategy are not arbitrary, contrary to traditional diagonal schemes (see, e.g., [175, 216, 240, 242]). They are members of a sequence of values depending on both the size of the initial hypercube $D = [a, b]$ and the number of executed subdivisions (see Theorem 3.2 in the previous chapter). In this way, the hyperintervals of the current partition $\{D^k\}$ form several groups. Each group is characterized by the length of the main diagonal of hyperintervals within the group. In the two-dimensional diagram (d_i, F_i), where d_i and F_i are from (4.5), the hyperintervals from a group are represented by dots with the same horizontal coordinate d_i. For example, in Fig. 4.2 there are seven different groups of hyperintervals with the horizontal coordinates equal to $d_A, d_B, d_C, d_E, d_F, d_G$, and d_H. Note that some groups of the current partition can be empty (see, e.g., the group with the horizontal coordinate between d_H and d_G in Fig. 4.2). These groups correspond to diagonals which are not present in the current partition but can be created (or were created) at the successive (previous) iterations of the algorithm.

It is easy to deduce (see [279]) from the scheme of the non-redundant partition strategy and Theorem 3.2 that there exists a correspondence between the length of the main diagonal of a hyperinterval $D_i \in \{D^k\}$ and a non-negative integer number. This number indicates how many partitions have been performed starting from the initial hypercube D to obtain the hyperinterval D_i. At each iteration $k \geq 1$ it can be considered as an index $l = l(k)$ of the corresponding group of hyperintervals having the same length of their main diagonals, where

$$0 \leq q(k) \leq l(k) \leq Q(k) < +\infty \tag{4.7}$$

and $q(k) = q$ and $Q(k) = Q$ are indices corresponding to the groups of the largest and smallest hyperintervals of $\{D^k\}$, respectively. When the algorithm starts, there exists only one hyperinterval—the admissible region D—which belongs to the group with the index $l = 0$. In this case, both indices q and Q are equal to zero. When a hyperinterval $D_i \in \{D^k\}$ from a group $l' = l'(k)$ is subdivided, all three generated hyperintervals are placed into the group with the index $l' + 1$. Thus, during the work of the algorithm, diagonals of hyperintervals become smaller and smaller, while the corresponding indices of groups of hyperintervals grow consecutively starting from zero.

For example, in Fig. 4.2 there are seven non-empty groups of hyperintervals of a partition $\{D^k\}$ and one empty group. The index $q(k)$ (index $Q(k)$) corresponds to the group of the largest (smallest) hyperintervals represented in Fig. 4.2 by dots with the horizontal coordinate equal to d_A (d_H). In Fig. 4.2 we have $Q(k) = q(k) + 7$. The empty group has the index $l(k) = Q(k) - 1$. Suppose that the hyperintervals D_A, D_H, and D_G (represented in Fig. 4.2 by the dots A, H, and G, respectively) will be subdivided at the kth iteration. In this case, the smallest index will remain the same, i.e., $q(k + 1) = q(k)$, since the group of the largest hyperintervals will not be empty, while the biggest index will increase, i.e., $Q(k + 1) = Q(k) + 1$, since a new group of the smallest hyperintervals will be created. The previously empty group $Q(k) - 1$ will be filled up by the new hyperintervals generated by partitioning the hyperinterval D_G and will have the index $l(k + 1) = Q(k + 1) - 2$.

4.2.3 Description of the MULTL Algorithm and its Convergence Analysis

Now we are ready to describe the MULTL algorithm. First, we present the algorithm and briefly comment upon it. Then we analyze its convergence properties.

MULTL is oriented on solving hard multidimensional multiextremal problems. To accomplish this task, a two-phase approach consisting of explicitly defined global and local phases is proposed. It is well known that DIRECT also balances global and local information during its work. However, its local phase is too pronounced in this balancing. As has already been mentioned at the beginning of this section, DIRECT executes too many function trials in regions of local optima and, therefore, manifests a slow convergence to the global minimizers if the objective function has many local minimizers.

In MULTL, when a sufficient number of subdivisions of hyperintervals near the current best point have been performed, the two-phase approach forces the algorithm to switch to the exploration of large hyperintervals that could contain better solutions. Since many subdivisions have been executed around the current best point, its neighborhood contains only small hyperintervals and large ones can only be located far from it. Thus, MULTL balances global and local search in a more sophisticated way trying to provide a faster convergence to the global minimizers of hard multiextremal functions.

The MULTL method consists of the following two phases: local improvement of the current estimate of the global minimum value (local phase) and examination of large unexplored hyperintervals in pursuit of new attraction regions of deeper local minimizers (global phase). Each of these phases can consist of several iterations. During the local phase the algorithm tries to better explore the subregion around the current best point. This phase finishes when the following two conditions are verified: (i) an improvement on at least 1% of the current estimate of the minimal function value is no longer achieved and (ii) a hyperinterval containing the current best point becomes the smallest one. After the end of the local phase the algorithm switches to the global phase.

The global phase consists of subdividing mainly large hyperintervals, located possibly far from the current best point. It is performed until a function value improving the current estimate of the global minimum value on at least 1% is obtained. When this happens, the algorithm switches to the local phase during which the obtained new solution is improved locally. During its work the algorithm can switch many times from the local phase to the global one. The algorithm stops when the number of generated trial points reaches the maximal allowed number.

Without loss of generality it is assumed that the admissible region $D = [a, b]$ in (3.2) is an N-dimensional hypercube. Suppose that at the iteration $k \geq 1$ of the algorithm a partition $\{D^k\}$ of $D = [a, b]$ has been obtained by applying the non-redundant diagonal partition scheme described in the previous chapter. Suppose also that each hyperinterval $D_i \in \{D^k\}$ from (3.4) is represented by a dot in the two-dimensional diagram (d_i, F_i), where d_i and F_i are from (4.5), and the groups of hyperintervals with the same length of their main diagonals are numerated by indices within a range from $q(k)$ up to $Q(k)$ from (4.7).

To describe MULTL formally, we need the following additional designations:

$f_{min}(k)$—the current estimate of the global minimum value of $f(x)$ found after $k - 1$ iterations (the term 'record' will be also used).

$x_{min}(k)$—coordinates of $f_{min}(k)$.

$D_{min}(k)$—the hyperinterval containing the point $x_{min}(k)$ (if $x_{min}(k)$ is a common vertex of several—up to 2^N—hyperintervals, then the smallest hyperinterval is considered).

f_{min}^{prec}—the old record. It serves to memorize the record $f_{min}(k)$ at the start of the current phase (local or global). The value of f_{min}^{prec} is updated when an improvement of the current record on at least 1% is obtained.

η—the parameter of the algorithm, $\eta \geq 0$. It prevents the algorithm from subdividing already well-explored small hyperintervals. If $D_t \in \{D^k\}$ is a non-dominated hyperinterval with respect to an estimate $\tilde{L}$ of the Lipschitz constant L, then this hyperinterval can be subdivided at the kth iteration only if the following condition is satisfied

$$R_t(\tilde{L}) \leq f_{min}(k) - \eta, \tag{4.8}$$

where the lower bound $R_t(\tilde{L})$ is calculated by formula (4.1). The value of η can be set in different ways which will be considered later during the discussion of the results of numerical experiments.

T_{max}—the maximal allowed number of trial points that the algorithm may generate. The algorithm stops when the number of generated trial points reaches T_{max}.

During the course of the algorithm the satisfaction of this termination criterion is verified after every subdivision of a hyperinterval.

Lcounter, Gcounter—the counters of iterations executed during the current local and global phases, respectively.

$p(k)$—the index of the group the hyperinterval $D_{min}(k)$ belongs to. Notice that the inequality $q(k) \leq p(k) \leq Q(k)$ is satisfied for any iteration number k. Since both local and global phases can embrace more than one iteration, the index $p(k)$ (as well as the indices $q(k)$ and $Q(k)$) can change (namely, increase) during these phases. Note also that the group $p(k)$ can be different from the groups containing hyperintervals with the smallest sum of the objective function values at their vertices (see two groups of hyperintervals represented in Fig. 4.2 by the horizontal coordinates equal to d_G and d_F). Moreover, the hyperinterval $D_{min}(k)$ is not represented necessarily by the 'lowest' point from the group $p(k)$ in the two-dimensional diagram (d_i, F_i)—even if the current estimate of the global minimum value of $f(x)$ is obtained at a vertex of $D_{min}(k)$, the function value at the other vertex can be too high and the sum of these two values can be greater than the corresponding value of another hyperinterval from the group $p(k)$.

p'—the index of the group containing the hyperinterval $D_{min}(k)$ at the start of the current phase (local or global). Hyperintervals from the groups with indices greater than p' are not considered when non-dominated hyperintervals are determined. Whereas the index $p(k)$ can assume different values during the current phase, the index p' remains, as a rule, invariable. It is changed only when it violates the left part of condition (4.7). This can happen when groups with the largest hyperintervals disappear and, therefore, the index $q(k)$ increases and becomes equal to p'. In this case, the index p' increases jointly with $q(k)$.

p''—the index of the group immediately preceding the group p', i.e., $p'' = p' - 1$. This index is used within the local phase and can increase if $q(k)$ increases during this phase.

r'—the index of the middle group of hyperintervals between the groups p' and $q(k)$, i.e., $r' = \lceil(q(k) + p')/2\rceil$. This index is used within the global phase as a separator between the groups of large and small hyperintervals. It can increase if $q(k)$ increases during this phase.

To clarify the introduced group indices, let us consider an example of a partition $\{D^k\}$ represented by the two-dimensional diagram in Fig. 4.2. Let us suppose that the index $q(k)$ of the group of the largest hyperintervals corresponding to the points with the horizontal coordinate d_A in Fig. 4.2 is equal to 10. The index $Q(k)$ of the group of the smallest hyperintervals with the main diagonals equal to d_H (see Fig. 4.2) is equal to $Q(k) = q(k) + 7 = 17$. Let us also assume that the hyperinterval $D_{min}(k)$ belongs to the group of hyperintervals with the main diagonals equal to d_G (see Fig. 4.2). In this case, the index $p(k)$ is equal to 15 and the index p' is equal to 15 too. The index $p'' = 15 - 1 = 14$ and it corresponds to the group of hyperintervals represented in Fig. 4.2 by the dots with the horizontal coordinate d_F. Finally, the index $r' = \lceil(10 + 15)/2\rceil = 13$ and it corresponds to hyperintervals with the main diagonals equal to d_E. The indices p', p'', and r' can change their values only if the index $q(k)$ increases. Otherwise, they remain invariable during the iterations of the current phase (local or global). At the same time, the index $p(k)$ can change at each iteration, as soon as a new estimate of the minimal function value belonging to a hyperinterval of a group different from $p(k)$ is obtained.

Now we are ready to present a formal scheme of the MULTL algorithm.

Step 1. *Initialization:* Set the current iteration number $k := 1$, the current record $f_{min}(k) := \min\{f(a), f(b)\}$, where a and b are from (3.2). Set group indices $q(k) := Q(k) := p(k) := 0$.

Step 2. *Local Phase:* Memorize the current record $f_{min}^{prec} := f_{min}(k)$ and perform the following steps:

Step 2.1. Set $Lcounter := 1$ and fix the group index $p' := p(k)$.

Step 2.2. Set $p'' := \max\{p' - 1, q(k)\}$.

Step 2.3. Determine non-dominated hyperintervals considering only groups of hyperintervals with the indices from $q(k)$ up to p''. Subdivide those non-dominated hyperintervals which satisfy inequality (4.8).
Set $k := k + 1$.

Step 2.4. Set $Lcounter := Lcounter + 1$ and check whether $Lcounter \leq N$. If this is the case, then go to **Step 2.2**. Otherwise, go to **Step 2.5**.

Step 2.5. Set $p' = \max\{p', q(k)\}$. Determine non-dominated hyperintervals considering only groups of hyperintervals with the indices from $q(k)$ up to p'. Subdivide those non-dominated hyperintervals which satisfy inequality (4.8).
Set $k := k + 1$.

Step 3. *Switch:* If condition

$$f_{min}(k) \leq f_{min}^{prec} - 0.01|f_{min}^{prec}| \tag{4.9}$$

is satisfied, then go to **Step 2** and repeat the local phase with the new obtained value of the record $f_{min}(k)$. Otherwise, if the hyperinterval $D_{min}(k)$ is not the smallest one, or the current partition of D consists only of hyperintervals with equal diagonals (i.e., $p(k) < Q(k)$ or $q(k) = Q(k)$), then go to **Step 2.1** and repeat the local phase with the old record f_{min}^{prec}.
If the obtained improvement of the current estimate of the minimal function value is not sufficient to satisfy (4.9) and $D_{min}(k)$ is the smallest hyperinterval of the current partition (i.e., all the following inequalities—(4.9), $p(k) < Q(k)$, and $q(k) = Q(k)$—fail), then go to **Step 4** and perform the global phase.

Step 4. *Global Phase:* Memorize the current record $f_{min}^{prec} := f_{min}(k)$ and perform the following steps:

Step 4.1. Set $Gcounter := 1$ and fix the group index $p' := p(k)$.

Step 4.2. Set $p' = \max\{p', q(k)\}$ and calculate the 'middle' group index $r' = \lceil (q(k) + p')/2 \rceil$.

Step 4.3. Determine non-dominated hyperintervals considering only groups of hyperintervals with the indices from $q(k)$ up to r'. Subdivide those non-dominated hyperintervals which satisfy inequality (4.8). Set $k := k + 1$.

Step 4.4. If condition (4.9) is satisfied, then go to **Step 2** and perform the local phase with the new obtained value of the record $f_{min}(k)$. Otherwise, go to **Step 4.5**.

Step 4.5. Set $Gcounter := Gcounter + 1$; check whether $Gcounter \leq 2^{N+1}$. If this is the case, then go to **Step 4.2**. Otherwise, go to **Step 4.6**.

Step 4.6. Set $p' = \max\{p', q(k)\}$. Determine non-dominated hyperintervals considering only groups of hyperintervals with the indices from $q(k)$ up to p'. Subdivide those non-dominated hyperintervals which satisfy inequality (4.8). Set $k := k + 1$.

Step 4.7. If condition (4.9) is satisfied, then go to Step 2 and perform the local phase with the new obtained value of the record $f_{min}(k)$. Otherwise, go to **Step 4.1**: update the value of the group index p' and repeat the global phase with the old record f_{min}^{prec}.

Let us give a few comments upon the introduced algorithm. It starts its work from the local phase. In the course of this phase, it subdivides non-dominated hyperintervals with the main diagonals greater than the main diagonal of $D_{min}(k)$ (i.e., from the groups with the indices from $q(k)$ up to p'; see Steps 2.1–2.4). This operation is repeated N times, where N is the problem dimension from (3.2). Recall that during each subdivision of a hyperinterval by the efficient diagonal partition scheme only one edge of the hyperinterval (namely, the longest edge given by formula (3.19)) is partitioned. Thus, performing N iterations of the local phase eventually subdivides all N sides of hyperintervals around the current best point. At the last, $(N + 1)$-th, iteration of the local phase (see Step 2.5) hyperintervals with the main diagonal equal to $D_{min}(k)$ is considered too. In this way, the hyperinterval containing the current best point can be partitioned too.

Thus, either the current record is improved or the hyperinterval providing this record becomes smaller. If the conditions of switching to the global phase (see Step 3) are not satisfied, the local phase is repeated. Otherwise, the algorithm switches to the global phase, thus avoiding unnecessary evaluations of $f(x)$ within already well-explored subregions.

During the global phase the algorithm searches for better new minimizers. It performs a series of loops (see Steps 4.1–4.7) while a non-trivial improvement of the current estimate of the minimal function value is not obtained, i.e., condition (4.9) is not satisfied. Within a loop of the global phase the algorithm performs a substantial number of subdivisions of large hyperintervals located far from the current best point, namely, hyperintervals from the groups with the indices from $q(k)$ up to r' (see Steps 4.2–4.5). Since each trial point can belong up to 2^N hyperintervals, the number of subdivisions should not be smaller than 2^N. The value of this number equal to 2^{N+1} has been chosen because it provided a good performance of the algorithm in our numerical experiments.

Note that the situation when the current estimate of the minimal function value is improved but the amount of this improvement is not sufficient to satisfy (4.9) can be verified at the end of a loop of the global phase (see Step 4.7). In this case, the algorithm is not switched to the local phase. It proceeds with the next loop of the global phase, eventually updating the index p' (see Step 4.1) but not updating the old record f_{min}^{prec}.

Let us now study convergence properties of MULTL during minimization of the function $f(x)$ from (3.1)–(3.3) with the maximal allowed number of generated trial points T_{max} is equal to infinity. In this case, the algorithm does not stop (the number of iterations k goes to infinity) and an infinite sequence of trial points $\{x^{j(k)}\}$ is generated. The following theorem establishes the so-called 'everywhere dense' convergence (i.e., convergence of the sequence of trial points to any point of the search domain) of MULTL.

Theorem 4.3 *For any point $x \in D$ and any $\delta > 0$ there exist an iteration number $k(\delta) \geq 1$ and a point $x' \in \{x^{j(k)}\}$, $k > k(\delta)$, such that $\|x - x'\| < \delta$.*

Proof Trial points generated by the MULTL method are vertices corresponding to the main diagonal of hyperintervals. Due to the non-redundant partition scheme described in Sect. 3.3, every subdivision of a hyperinterval produces three new hyperintervals with the volume equal to a third of the volume of the subdivided hyperinterval and the proportionally smaller main diagonals. Thus, once fixed a positive value of δ, it is sufficient to prove that after a finite number of iterations $k(\delta)$ the largest hyperinterval of the current partition of D will have the main diagonal smaller than δ. In such a case, in δ-neighborhood of any point of D there will exist at least one trial point generated by the algorithm.

To see this, let us fix an iteration number k' and consider the group $q(k')$ of the largest hyperintervals of a partition $\{D^{k'}\}$. As can be seen from the scheme of the algorithm, for any $k' \geq 1$ this group is taken into account when non-dominated hyperintervals are determined. Moreover, a hyperinterval $D_t \in \{D^{k'}\}$ from this group having the smallest sum of the objective function values at its vertices is partitioned at each iteration $k \geq 1$ of the algorithm. This happens because there always exists a sufficiently large estimate L_∞ of the Lipschitz constant L such that the hyperinterval D_t is a non-dominated hyperinterval with respect to L_∞ and condition (4.8) is satisfied for the lower bound $R_t(L_\infty)$ (see Fig. 4.2). Three new hyperintervals generated during the subdivision of D_t by using the non-redundant diagonal partition strategy are inserted into the group with the index $q(k') + 1$. Hyperintervals of the group $q(k') + 1$ have the volume equal to a third of the volume of hyperintervals of the group $q(k')$.

Since each group contains only a finite number of hyperintervals, after a sufficiently large number of iterations $k > k'$ all hyperintervals of the group $q(k')$ will be subdivided. The group $q(k')$ will become empty and the index of the group of the largest hyperintervals will increase, i.e., $q(k) = q(k') + 1$. Such a procedure will be repeated with a new group of the largest hyperintervals. So, when the number of iterations grows up, the index $q(k)$ increases and due to (4.7) the index $Q(k)$

increases, too. This means that there exists a finite number of iterations $k(\delta)$ such that after performing $k(\delta)$ iterations of the algorithm the largest hyperinterval of the current partition $\{D^{k(\delta)}\}$ will have the main diagonal smaller than δ. The theorem has been proved. $\qquad\square$

4.3 One-Point-Based Method MULTK for Differentiable Problems

In this section, another example of applying the non-redundant partition scheme to the construction of efficient Lipschitz global optimization algorithms is presented. In particular, the class of problems with differentiable objective functions having the Lipschitz gradients $f'(x) := \nabla f(x)$ is considered, i.e.,

$$f^* = f(x^*) = \min_{x \in D} f(x), \qquad (4.10)$$

$$\|f'(x') - f'(x'')\| \le K\|x' - x''\|, \quad x', x'' \in D, \quad 0 < K < \infty, \qquad (4.11)$$

where

$$D = [a, b] = \{x \in \mathbb{R}^N : a(j) \le x(j) \le b(j),\, 1 \le j \le N\}, \quad a, b \in \mathbb{R}^N. \quad (4.12)$$

It is supposed in this formulation that the objective function $f(x)$ can be black-box, multiextremal, its gradient $f'(x) = \left(\frac{\partial f(x)}{\partial x(1)}, \frac{\partial f(x)}{\partial x(2)}, \dots, \frac{\partial f(x)}{\partial x(N)}\right)^T$ (which could be itself an expensive black-box vector-function) can be calculated during the search, and $f'(x)$ is Lipschitz-continuous with some fixed, but unknown, constant $K, 0 < K < \infty$, over D. These problems are often encountered in engineering applications (see, e.g., [242, 290, 323]), particularly, in electrical engineering optimization problems (see, e.g., [282, 290, 323]).

In the literature, several methods for solving this problem have been proposed. As in the case of problem (3.1), (3.2), they can be also distinguished, for instance, with respect to the way the Lipschitz constant K is estimated in their work. There exist algorithms using an a priori given estimate of K (see, e.g., [13, 34, 276]), its adaptive estimates (see, e.g., [117, 276, 290]), and adaptive estimates of local Lipschitz constants (see, e.g., [276, 290]). Algorithms working with a number of Lipschitz constants for $f'(x)$ chosen from a set of possible values varying from zero to infinity were not known till 2009 when such an algorithm for solving the one-dimensional problem (2.2), (2.4) has been proposed in [177] (see Chap. 2). Its extension to the multidimensional case is not a trivial task in contrast to the DIRECT method (see [154]) proposed in 1993 for solving problems with the Lipschitz objective function.

In the paper [179], this more than 15-year open problem of constructing multidimensional global optimization methods working with multiple estimates of the

Lipschitz constants for $f'(x)$ has been successfully solved. Following this paper, the multidimensional geometric method MULTK for finding solutions to the problem (4.10)–(4.12) is described and discussed here. It uses the non-redundant one-point-based partitioning strategy (see [280, 290] and Sect. 3.3) and works with a number of estimates of the Lipschitz constant K for $f'(x)$. Such multiple (from zero to infinity) estimates of K from (4.11) are used to calculate the lower bounds of the objective function over hyperintervals of a current partition of the search domain and to produce new trial points (in this case, each trial consists of evaluating both the objective function $f(x)$ and its gradient $f'(x)$). Recall that in the framework of geometric algorithms this kind of estimates of the Lipschitz constant can be interpreted as an examination of all admissible minorant functions that could be built during each current iteration of the algorithm without constructing a specific one. A particular attention in this algorithm, as well as in the MULTL method from the previous section, is given to the improvement of the current minimal function value (*record value*) in order to provide a faster convergence to a global minimizer. As demonstrated by extensive numerical experiments, the usage of gradients allows one to obtain, as expected, an acceleration in comparison with the DIRECT-based methods.

This section is organized as follows. A theoretical background of the MULTK algorithm is first presented. Then, a formal description of the algorithm and its convergence analysis are reported. Results of numerical experiments are given in the next section, together with those of the MULTL method.

4.3.1 Theoretical Background of MULTK: *Lower Bounds*

Let us consider an iteration $k \geq 1$ of the MULTK algorithm and a current partition $\{D^k\}$ of the search hyperinterval $D = [a, b]$ into hyperintervals $D_i = [a_i, b_i]$, $1 \leq i \leq m(k)$ generated by the non-redundant one-point-based partition strategy as described in Sect. 3.3. Over these hyperintervals the values of both the function and its gradient are obtained (evaluated or read from the vertex database) at trial points $x^{j(k)} = a_i, j(k) \geq 1$. As usual, in order to choose some hyperintervals for the further partition, the hyperintervals characteristics should be calculated and an estimate of the lower bound of $f(x)$ over a hyperinterval is one of the possible characteristics of this hyperinterval. The following result holds.

Theorem 4.4 *Let $\tilde{K}$ be an estimate of the Lipschitz constant K for $f'(x)$ from (4.11), $\tilde{K} \geq K$ and $D_i = [a_i, b_i]$ be a hyperinterval of a current partition $\{D^k\}$ with a trial point a_i. Then, a value $R_i(\tilde{K})$ of the characteristic of D_i can be found such that it is the lower bound of $f(x)$ over D_i, i.e., $R_i(\tilde{K}) \leq f(x)$, $x \in D_i$.*

Proof Let us prove the theorem in a constructive way. It is known (see, e.g., [87, 221, 223]) that for a differentiable function $f(x)$ over a hyperinterval $D_i = [a_i, b_i]$ the following inequality is satisfied:

$$f(x) \geq Q(x, \tilde{K}), \quad x \in D_i, \tag{4.13}$$

where the quadratic minorant function $Q(x, \tilde{K})$ is defined over D_i as

$$Q(x, \tilde{K}) = f(a_i) + \langle f'(a_i), (x - a_i) \rangle - 0.5\tilde{K}\|x - a_i\|^2, \quad x \in D_i. \tag{4.14}$$

Here $\langle \cdot, \cdot \rangle$ is the scalar product, $\| \cdot \|$ is the Euclidean norm in R^N, and

$$g(x) = f(a_i) - \langle f'(a_i), (x - a_i) \rangle$$

is the linear approximation of $f(x)$ over D_i.

From inequality (4.13) the following estimates can be obtained:

$$f(x) \geq f(a_i) + \langle f'(a_i), (x - a_i) \rangle - 0.5\tilde{K}\|b_i - a_i\|^2 \geq$$

$$F_i - 0.5\tilde{K}\|b_i - a_i\|^2, \quad x \in D_i,$$

where F_i is the minimum value of the linear approximation $g(x)$ over D_i, i.e.,

$$F_i = f(a_i) + \min_{x \in D_i}\langle f'(a_i), (x - a_i) \rangle. \tag{4.15}$$

Since the function $g(x)$ is linear, its minimum (4.15) is obtained in the vertex z_i of the hyperinterval $D_i = [a_i, b_i]$ whose coordinates $z_i(j), j = 1, \ldots, N$ can be calculated as follows:

$$z_i(j) = \begin{cases} a_i(j), & \text{if either } b_i(j) > a_i(j) \text{ and } \frac{\partial f(a_i)}{\partial x(j)} \geq 0, \\ & \text{or } b_i(j) < a_i(j) \text{ and } \frac{\partial f(a_i)}{\partial x(j)} < 0; \\ b_i(j), & \text{if either } b_i(j) > a_i(j) \text{ and } \frac{\partial f(a_i)}{\partial x(j)} < 0, \\ & \text{or } b_i(j) < a_i(j) \text{ and } \frac{\partial f(a_i)}{\partial x(j)} \geq 0. \end{cases} \tag{4.16}$$

The corresponding value F_i from (4.15) is therefore equal to

$$F_i = f(a_i) + \langle f'(a_i), (z_i - a_i) \rangle. \tag{4.17}$$

It is clear now that the value

$$R_i = R_i(\tilde{K}) = F_i - 0.5\tilde{K}\|b_i - a_i\|^2 \tag{4.18}$$

satisfies the inequality

$$R_i \leq f(x), \quad x \in D_i,$$

and, therefore, it can be taken as the characteristic value of D_i that estimates the lower bound of $f(x)$ over D_i. The theorem has been proved.   □

Clearly, analogous results can be obtained in the case of hyperintervals D_i with trial points b_i rather than a_i.

In Fig. 4.3, a quadratic minorant function $Q(x, \tilde{K})$ from (4.14) is illustrated for $f(x)$ over a hyperinterval D_i. Here, the characteristic value R_i coincides with the minimum value of $Q(x, \tilde{K})$ obtained at the point b_i of the main diagonal of D_i. In

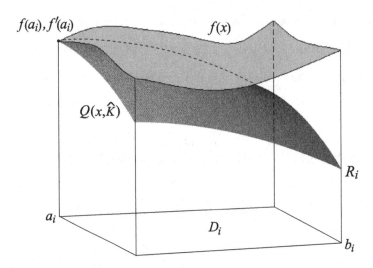

Fig. 4.3 A quadratic minorant function $Q(x, \tilde{K})$ for $f(x)$ over a hyperinterval $D_i = [a_i, b_i]$

general, as it can be seen from (4.14), the value R_i is smaller than or equal to the minimum value of $Q(x, \tilde{K})$ over D_i.

4.3.2 Theoretical Background of MULTK: Non-dominated Hyperintervals

By using the obtained characteristics of hyperintervals, the relation of domination can be established between every two hyperintervals of a current partition $\{D^k\}$ of D and a set of non-dominated hyperintervals can be identified for a possible subdivision at the current iteration of the MULTK algorithm (see [177, 289]). This can be done in a way similar to that of the MULTL method from the previous section. However, there exists an important difference in this context that lies in the graphical representation of the hyperintervals.

Let us consider this important issue more in detail and generalize to the multi-dimensional case the approach proposed by the authors in [177] (see Chap. 2) for one-dimensional problems. In particular, as shown in [179], both a multidimensional interval $D_i = [a_i, b_i]$ of a current partition $\{D^k\}$ and the respective characteristic R_i using the gradient can be represented in a two-dimensional diagram similar to those proposed in [154, 289] for derivative-free methods. Difficulties in the construction of such a diagram were among the main reasons that prevented people to propose methods using several estimates of K in their work (see Chap. 2 for a detailed discussion upon this issue in the simplest one-dimensional case).

In order to construct such a diagram, we take for the dot, corresponding to a hyperinterval D_i, the vertical coordinate F_i from (4.15)–(4.17) and the horizontal coordinate d_i equal to half of the squared length of the main diagonal of D_i, i.e.,

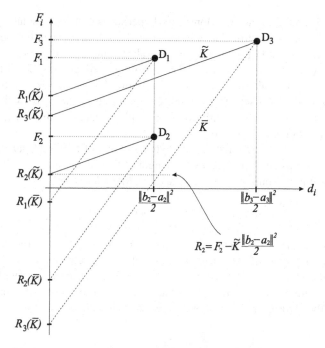

Fig. 4.4 Graphical representation of hyperintervals in MULTK

$$d_i = 0.5\|b_i - a_i\|^2.$$

For example, in Fig. 4.4, a partition of the search domain D consisting of three hyperintervals is represented by the dots D_1, D_2, and D_3. Let us suppose that the Lipschitz constant K for the gradient $f'(x)$ is estimated by $\tilde{K}$, $\tilde{K} \geq K$. The characteristic R_i of a hyperinterval D_i, $i = 1, 2, 3$, can be graphically obtained as the vertical coordinate of the intersection point of the line passed through the point D_i with the slope $\tilde{K}$ and the vertical coordinate axis (see Fig. 4.4). It is easy to see, that with respect to the estimate $\tilde{K}$ the hyperinterval D_2 dominates both hyperintervals D_1 and D_3 and the hyperinterval D_3 dominates D_1.

If a higher estimate $\bar{K} > \tilde{K}$ of the Lipschitz constant K is considered (see Fig. 4.4), the hyperinterval D_2 still dominates D_1 with respect to $\bar{K}$, because $R_2(\bar{K}) < R_1(\bar{K})$. But D_2 in its turn is dominated by the hyperinterval D_3 with respect to $\bar{K}$, because $R_2(\bar{K}) > R_3(\bar{K})$ (see Fig. 4.4).

Since the exact Lipschitz constant K for $f'(x)$ (or its valid overestimate) is unknown in the stated problem too, the following definition can be useful.

Definition 4.4 A hyperinterval $D_t \in \{D^k\}$ is called *non-dominated* if there exists an estimate $0 < \tilde{K} < \infty$ of the Lipschitz constant K such that D_t is non-dominated with respect to $\tilde{K}$, in the sense of definitions from the previous sections.

This definition means that non-dominated hyperintervals have the smallest characteristics (4.18) for some particular estimate of the Lipschitz constant for the gradient $f'(x)$. For example, in Fig. 4.4 the hyperintervals D_2 and D_3 are non-dominated.

It can be demonstrated following the reasoning used in [177, 289] that non-dominated hyperintervals (in the sense of Definition 4.4) are located on the lower-right convex hull of the set of dots representing the hyperintervals of the current partition of D and can be efficiently found by applying an algorithm for identifying the convex hull of the dots (see, e.g., [141, 154, 290]).

Similarly to the MULTL method, the hyperintervals of a current partition of D form several groups characterized by the length of their main diagonals. Therefore, the same system of group indices can be used in MULTK, as well:

$$0 \le q_\infty(k) \le s(k) \le q_0(k) < +\infty, \qquad (4.19)$$

where $q_\infty(k)$ and $q_0(k)$ are indices corresponding to the groups of the largest and smallest hyperintervals of the current partition of D, respectively.

Once a non-dominated hyperinterval $D_t = [a_t, b_t]$ is determined (with respect to some estimate $\tilde{K}$ of the Lipschitz constant K), it can be subdivided at the next iteration of the algorithm if the following condition is satisfied:

$$R_t(\tilde{K}) \le f_{\min}(k) - \eta, \qquad (4.20)$$

where R_t is calculated by (4.18), $f_{\min}(k)$ is the record value, i.e., the current minimal function value (attained at the record point $x_{\min}(k)$), and η is the parameter of the algorithm, $\eta \ge 0$. Remember that condition (4.20) prevents the algorithm from subdividing already well-explored small hyperintervals.

It should be mentioned in this occasion that, together with non-dominated hyperintervals, a hyperinterval $D_{\min}(k) = [a_{\min}, b_{\min}]$ containing the record point (called hereafter the record hyperinterval) is also considered for a possible partition during the work of the algorithm. Among different hyperintervals the record point $x_{\min}(k)$ can belong to (up to 2^N), the record hyperinterval is that with the smallest characteristic and can be changed during subdivisions. Hereinafter, the index of the group the hyperinterval $D_{\min}(k)$ belongs to will be indicated as $p(k)$. During the work of the algorithm the satisfaction of inequalities (4.19) is ensured for this index which can be eventually updated together with $q_0(k)$ and $q_\infty(k)$ (see [289] for details).

4.3.3 Description of the MULTK Algorithm and its Convergence Analysis

Let us now present the computational scheme of MULTK and analyze its convergence properties.

As the previously described MULTL method, the MULTK algorithm consists of the following explicitly defined phases: (1) an exploration (global) phase, at which an examination of large hyperintervals (possibly located far away from the record point) is performed in order to capture new subregions with better function values; (2) a record improvement (local) phase, at which the algorithm tries to better inspect the subregion around the record point.

The exploration phase consists of several iterations (namely, $N + 1$ where N is the problem dimension), each serves for determining non-dominated hyperintervals and partitioning them. The number $N + 1$ has been chosen since each subdivision of a hyperinterval by the non-redundant one-point-based scheme is performed perpendicularly to only one edge of the hyperinterval (to the longest edge from (3.19)), thus, the number of iterations within a phase of the algorithm should be correlated with the hyperintervals dimension.

This phase is interrupted after finishing an iteration if an improvement on at least 1% of the minimal function value is reached, i.e., if

$$f_{\min}(k) \leq f_{\min}^{\text{prec}} - 0.01 |f_{\min}^{\text{prec}}|, \tag{4.21}$$

where $f_{\min}^{\text{prec}}$ is the record value memorized at the start of the exploration phase.

Condition (4.21) is verified after each iteration of the exploration phase and is used to switch the algorithm to the record improvement phase. This local phase is also launched when the exploration phase finishes without having improved the record value, but only if the record hyperinterval $D_{\min}(k)$ is not the smallest one within the current partition of hyperintervals. Otherwise, the algorithm re-initiates another global exploration phase without forcing the local one.

The record improvement phase, at a single iteration, performs several subdivisions (namely, N) of the record hyperinterval trying to improve the record value. During this process a new record value can appear. In this case, a new record hyperinterval can be considered for remaining subdivisions.

The record hyperinterval subdivisions are also performed by means of the non-redundant one-point-based strategy. Of course, other possible local improvement techniques can be used for this scope (see, e.g., [112, 221, 223]) but in this case the resulting trial points can be managed with a difficulty within the vertex database mentioned in Sect. 3.3.

It should be stressed that the available gradient information allows us to terminate automatically the record improvement phase. In fact, the record hyperinterval is not further subdivided if the gradient projection on the directions parallel to the record hyperinterval sides becomes non-negative, i.e., when the following condition is satisfied:

$$\frac{\partial f(a_{\min})}{\partial x(j)} (b_{\min}(j) - a_{\min}(j)) \geq 0, \quad 1 \leq j \leq N. \tag{4.22}$$

Either in this case or when the prefixed number N of subdivisions are normally performed (without meeting conditions (4.22)), the algorithm switches again to the global exploration phase and continues its work.

The algorithm stops when the number of generated trial points reaches the maximal allowed number $P_{\max}$. The satisfaction of this termination criterion is checked after every subdivision of a hyperinterval. The current record value $f_{\min}$ and the current record point $x_{\min}$ can be taken as approximations of the global minimum value f^* and the global minimizer x^* from (4.10), respectively.

A formal description of the MULTK algorithm follows below (it is assumed without loss of generality that the admissible region $D = [a, b]$ in (4.12) is an N-dimensional hypercube).

Step 0 (*Initialization*). Set the iteration counter $k := 1$. Let the first evaluation of $f(x)$ and $f'(x)$ be performed at the vertex a of the initial hyperinterval $D = [a, b]$, i.e., $x^1 := a$. Set the current partition of the search interval as $D^1 := \{[a_1, b_1]\}$, where $a_1 = a$, $b_1 = b$, and the current number of hyperintervals $m(1) := 1$. Set $f_{\min}(1) := f(x^1)$, $x_{\min}(1) := a$, and $D_{\min}(1) := [a_1, b_1]$. Set group indices $q_\infty(1) := q_0(1) := p(1) := 0$.

Suppose now that $k \geq 1$ iterations of the algorithm have already been executed. The next iterations of the algorithm consist of the following steps.

Step 1 (*Exploration Phase*). Memorize the current record $f_{\min}^{\text{prec}} := f_{\min}(k)$, set the counter of iterations during the exploration phase $k_g := 1$ and perform the following steps:

Step 1.1. Identify the set of non-dominated hyperintervals considering only groups of large hyperintervals (namely, those with the current indices from $q_\infty(k)$ up to $\lceil (q_\infty(k) + p(k))/2 \rceil$). Subdivide the non-dominated hyperintervals which satisfy inequality (4.20) and produce new trial points (or read the existing ones from the vertex database) according to Sect. 3.3. Set $k := k + 1$ and update hyperintervals indices if necessary (see [289] for details).

Step 1.2. If condition (4.21) is satisfied, then go to Step 2 and execute the record improvement phase. Otherwise, go to Step 1.3.

Step 1.3. Increase the counter $k_g := k_g + 1$: check whether $k_g \leq N$. If this is the case, then go to Step 1.1 (continue the exploration of large hyperintervals). Otherwise, go to Step 1.4 (perform the final iteration of the exploration phase by considering more groups of hyperintervals).

Step 1.4. Identify the set of non-dominated hyperintervals considering the current groups of hyperintervals from $q_\infty(k)$ up to $p(k)$. Subdivide the non-dominated hyperintervals which satisfy inequality (4.20) and produce new trial points (or read the existing ones from the vertex database) according to Sect. 3.3. Set $k := k + 1$, update all necessary indices.

Step 1.5. If the record hyperinterval is not the smallest one, i.e., if $p(k) < q_0(k)$, then then go to Step 2 and execute the record improvement phase. Otherwise, go to Step 1 and repeat the exploration phase updating the value $f_{\min}^{\text{prec}}$.

Step 2 (*Record Improvement Phase*). Set $k := k + 1$. Set the counter of iterations during the record improvement phase $k_l := 1$ and perform the following steps:

Step 2.1. Subdivide the record hyperinterval $D_{\min}(k)$ and produce a new trial point (or read the existing one from the vertex database) according to Sect. 3.3.

Update hyperintervals indices and the record hyperinterval index if necessary.

Step 2.2. Increase the counter $k_l := k_l + 1$: check whether $k_l \leq N$. If this is the case, then go to Step 1 (perform a new exploration of large hyperintervals). Otherwise, go to Step 2.1 (continue the local exploration in the neighborhood of the record point).

Let us now study convergence properties of the MULTK method during minimization of the function $f(x)$ from (4.10)–(4.12) when the maximal allowed number of generated trial points P_{max} is equal to infinity. In this case, the algorithm does not stop (the number of iterations k goes to infinity) and an infinite sequence of trial points $\{x^{j(k)}\}$ is generated.

Theorem 4.5 *The MULTK algorithm manifests the everywhere dense convergence.*

Proof The result is proved similarly to the proof of Theorem 4.3. In this context, it is sufficient to add the following remark. Remember that the record hyperinterval D_{min} is itself represented by a dot in the two-dimensional diagram of the current partition. This hyperinterval can be subdivided either separately during the record improvement phase or as a non-dominated hyperinterval during the exploration phase at which the satisfaction of condition (4.22) is not taken in consideration. □

To conclude the theoretical description of the MULTK algorithm we would like to highlight that the usage of all possible estimates of the Lipschitz constant in its work leads to the convergence of the everywhere dense type. If the Lipschitz constant L (or its valid estimate) of the objective function $f(x)$ or the Lipschitz constant K (or its valid estimate) of the gradient $f'(x)$ can be used by a global optimization method, other types of convergence can be established for such an algorithm (see, e.g., methods from [150, 242, 290, 291, 323]).

4.4 Numerical Experiments with the MULTL and MULTK Methods

In this section, we present results of some numerical experiments performed to compare the MULTL and MULTK methods with two methods belonging to the same class with respect to the estimation of the Lipschitz constants: the original DIRECT algorithm from [154] and its locally biased modification DIRECTl from [106, 107]. The implementation of these two methods described, e.g., in [106] and downloadable from http://www4.ncsu.edu/~ctk/SOFTWARE/DIRECTv204.tar.gz has been used in all the experiments.

To execute a numerical comparison, we need to define the parameter η of both the algorithms in (4.8) and (4.20). This parameter can be set either independently of the current record $f_{min}(k)$ or in a relation with it. Since the objective function $f(x)$ is supposed to be black-box, it is not possible to know a priori which of these two ways is preferable.

In DIRECT [154], where a similar parameter is used, a value η related to the current estimate $f_{\min}(k)$ of the minimal function value is fixed as follows:

$$\eta = \varepsilon |f_{\min}(k)|, \quad \varepsilon \geq 0. \tag{4.23}$$

The choice of ε between 10^{-3} and 10^{-7} has demonstrated good results for DIRECT on a set of test functions (see [154]). Later, formula (4.23) has been used by many authors (see, e.g., [44, 93, 106, 107, 141]) and has been also realized in the implementation of DIRECT (see, e.g., [106]) taken for a numerical comparison with the described methods. Since the value of $\varepsilon = 10^{-4}$ recommended in [154] has produced the most robust results for DIRECT (see, e.g., [106, 107, 141, 154]), exactly this value was used in (4.23) for DIRECT in our numerical experiments. In order to have comparable results, the same formula (4.23) and $\varepsilon = 10^{-4}$ were used in the MULTL and MULTK methods, as well.

The global minimizer $x^* \in D$ was considered to be found when an algorithm generated a trial point x' inside a hyperinterval with a vertex x^* and the volume smaller than the volume of the initial hyperinterval $D = [a, b]$ multiplied by an accuracy coefficient Δ, $0 < \Delta \leq 1$, i.e.,

$$|x'(i) - x^*(i)| \leq \sqrt[N]{\Delta}(b(i) - a(i)), \quad 1 \leq i \leq N, \tag{4.24}$$

where N is the problem dimension. This condition means that, given Δ, a point x' satisfies (4.24) if the hyperinterval with the main diagonal $[x', x^*]$ and the sides proportional to the sides of the initial hyperinterval $D = [a, b]$ has a volume at least Δ^{-1} times smaller than the volume of D. Note that if in (4.24) the value of Δ is fixed and the problem dimension N increases, the length of the diagonal of the hyperinterval $[x', x^*]$ increases, too. In order to avoid this undesirable growth, the value of Δ was progressively decreased when the problem dimension increased.

We stopped the algorithms either when the maximal number of trials $T_{\max}$ was reached or when condition (4.24) was satisfied. Note that such a stopping criterion is acceptable only when the global minimizer x^* is known, i.e., for the case of test functions. When a real black-box objective function is minimized and global minimization algorithms have an internal stopping criterion, they execute a number of iterations (that can be very high) after a 'good' estimate of f^* has been obtained in order to demonstrate a 'goodness' of the found solution (see, e.g., [148, 242, 323]).

In the numerical experiments presented here, the GKLS-generator from [111] is used. It generates classes of multidimensional and multiextremal test functions with known local and global minima. The procedure of generation consists of defining a convex quadratic function (paraboloid) systematically distorted by polynomials. Each test class provided by the generator includes 100 functions and is defined only by the following five parameters:

N—the problem dimension;

m—the number of local minimizers;

f^*—the value of the global minimum;

ρ^*—the radius of the attraction region of the global minimizer;

r^*—the distance from the global minimizer to the vertex of the paraboloid.

The other necessary parameters are chosen randomly by the generator for each test function of the class. Note that the generator always produces the same test classes for a given set of the user-defined parameters allowing one to perform repeatable numerical experiments.

It should be emphasized that, by changing the user-defined parameters, classes with different properties can be created. For example, given a fixed dimension of the functions and the desired number of local minimizers, a more difficult class can be created either by shrinking the attraction region of the global minimizer or by moving the global minimizer far away from the paraboloid vertex.

For conducting numerical experiments, we used eight GKLS classes of continuously differentiable test functions of dimensions $N = 2, 3, 4,$ and 5. The number of local minimizers m chosen was equal to 10 and the global minimum value f^* was fixed equal to -1.0 for all classes (these values are default settings of the generator). For each particular problem dimension N we considered two test classes: a simple class and a hard one. The difficulty of a class was increased either by decreasing the radius ρ^* of the attraction region of the global minimizer (as for two- and five-dimensional classes), or by increasing the distance r^* from the global minimizer x^* to the paraboloid vertex P (three- and four-dimensional classes). The values of the parameters r^* and ρ^* used in the experiments are indicated in Table 4.1 (the third and the fourth columns, respectively), together with the problem dimension N (the first column) and the relative difficulty of the corresponding class (the second column, designated by the symbol $\mathscr{D}$, where the label 'Simple' corresponds to a simple class and the label 'Hard' corresponds to a more difficult class). Finally, the accuracy coefficient Δ from (4.24) is given in the last column in Table 4.1.

In Fig. 4.5, an example of a test function from the following continuously differentiable GKLS class is given: $N = 2, m = 10, f^* = -1, \rho^* = 0.10,$ and $r^* = 0.90$ (the parameters $N, r^*,$ and ρ^* of this class are given in Table 4.1 on the second line). The function is defined over the region $D = [-1, 1]^2$ and its number is 87 in the given test class. The randomly generated global minimizer of this func-

Table 4.1 Eight GKLS classes of test functions used in numerical experiments with the MULTL and MULTK methods: for each problem dimension N a 'simple' and a 'hard' classes are considered

N	$\mathscr{D}$	r^*	ρ^*	Δ
2	Simple	0.90	0.20	10^{-4}
2	Hard	0.90	0.10	10^{-4}
3	Simple	0.66	0.20	10^{-6}
3	Hard	0.90	0.20	10^{-6}
4	Simple	0.66	0.20	10^{-6}
4	Hard	0.90	0.20	10^{-6}
5	Simple	0.66	0.30	10^{-7}
5	Hard	0.66	0.20	10^{-7}

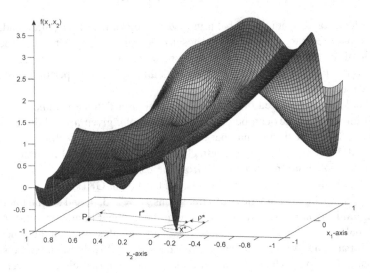

Fig. 4.5 An example of a two-dimensional function from the GKLS test class; by changing the user-defined parameters ρ^* (the radius of the attraction region of the global minimizer x^*) and/or r^* (the distance from the global minimizer to the vertex P of the paraboloid), more or less difficult classes can be created

tion is $x^* = (-0.767, -0.076)$ and the coordinates of the paraboloid vertex P are $(-0.489, 0.780)$.

We stopped algorithms either when the maximal number of trials T_{max} equal to $1\,000\,000$ was reached, or when condition (4.24) was satisfied. To describe experiments, we introduce the following designations:

T_s—the number of trials performed by the method under consideration to solve the problem number s, $1 \leq s \leq 100$, of a fixed test class. If the method was not able to solve a problem j in fewer than T_{max} function evaluations, T_j equal to T_{max} was taken.

M_s—the number of hyperintervals generated to solve the problem s.

The following four criteria were used to compare the methods.

Criterion C1. Number of trials T_{s^*} required for a method to satisfy condition (4.24) for *all* 100 functions of a particular test class, i.e.,

$$T_{s^*} = \max_{1 \leq s \leq 100} T_s, \quad s^* = \arg \max_{1 \leq s \leq 100} T_s. \tag{4.25}$$

Criterion C2. The corresponding number of hyperintervals, M_{s^*}, generated by the method, where s^* is from (4.25).

Criterion C3. Average number of trials T_{avg} performed by the method during minimization of *all* 100 functions from a particular test class, i.e.,

$$T_{avg} = \frac{1}{100} \sum_{s=1}^{100} T_s.$$ (4.26)

Criterion C4. Number p (number q) of functions from a class for which DIRECT or DIRECTl executed less (more) function evaluations than the MULTL or the MULTK methods. If T_s is the number of trials performed by the MULTL or by the MULTK methods and T_s' is the corresponding number of trials performed by a competing method, p and q are evaluated as follows

$$p = \sum_{s=1}^{100} \delta_s', \quad \delta_s' = \begin{cases} 1, & T_s' < T_s, \\ 0 & \text{otherwise}; \end{cases}$$ (4.27)

$$q = \sum_{s=1}^{100} \delta_s, \quad \delta_s = \begin{cases} 1, & T_s < T_s', \\ 0 & \text{otherwise}. \end{cases}$$ (4.28)

If $p + q < 100$, then both the methods under consideration solve the remaining $100 - p - q$ problems with the same number of function evaluations.

Note that the results based on Criteria C1 and C2 are mainly influenced by minimization of the most difficult functions of a class. Criteria C3 and C4 deal with average data of a class.

Criterion C1 is of fundamental importance for the methods comparison on the whole test class because it shows how many trials it is necessary to execute to solve *all* the problems of a class. Thus, it represents the worst case scenario of the given method on the fixed class.

At the same time, the number of generated hyperintervals (Criterion C2) provides an important characteristic of any partition algorithm for solving the global optimization problem. It reflects indirectly the degree of a qualitative examination of D during the search for the global minimum. The greater the number, the more information about the admissible domain is available and, therefore, the smaller the risk should be of missing the global minimizer. However, algorithms should not generate many redundant hyperintervals since this slows down the search and is, therefore, a disadvantage of the method.

Let us first compare the methods on Criteria C1 and C2, giving the results of the eight GKLS test classes for both the methods MULTL and MULTK. Recall that the accuracy coefficient Δ from (4.24), used in the experiments with each of the eight GKLS test classes, is given in the last column in Table 4.1.

Table 4.2 reports the maximal number of trials required for satisfying condition (4.24) for half of the functions of a particular class (columns '50%') and for all 100 function of the class (columns '100%'), for all the considered methods. The notation '$> 10^6$ (j)' means that after 1 000 000 function evaluations the method under consideration was not able to solve j problems. The corresponding numbers of generated hyperintervals (for all the methods) are indicated in Table 4.3. Since DIRECT and DIRECTl use during their work the center-sampling partition strategy, the number of generated trial points and the number of generated hyperintervals coincide for these methods.

Notice that on half of the test functions from each class (which were the most simple for each method with respect to the other functions of the class) the described methods manifested a good performance with respect to DIRECT and DIRECT*l* in terms of the number of generated trial points (see columns '50%' in Table 4.2). When all the functions were taken in consideration (and, consequently, difficult functions of the class were considered too), the number of trials produced by the MULT*L* and MULT*K* methods was much fewer in comparison with two other methods (see columns '100%' in Table 4.2), ensuring at the same time a substantial examination of the admissible domain (see Table 4.3).

Table 4.4 summarizes (based on the data from Table 4.2) the results (in terms of Criterion C1) of numerical experiments performed on 800 test functions from GKLS continuously differentiable classes. It represents the ratio between the maximal number of trials performed by DIRECT and DIRECT*l* with respect to the corresponding number of trials performed by the MULT*L* or by the MULT*K* methods. It can be seen from Table 4.4 that the methods based on the non-redundant partition strategy outperform both competitors significantly on the given test classes when Criteria C1 and C2 are considered.

Let us now compare the methods using Criteria C3 and C4. Tables 4.5 and 4.6 report the average number of trials performed during minimization of all 100 GKLS functions from the same class, when the methods MULT*L* and MULT*K* are used, respectively (Criterion C3). The 'Improvement' columns in these tables represent the ratios between the average numbers of trials performed by DIRECT and DIRECT*l* with respect to the corresponding numbers of trials performed by the MULT*L* or by the MULT*K* algorithms. The symbol '>' reflects the situation when not all functions of a class were successfully minimized by the method under consideration in the sense of condition (4.24). This means that the method stopped when T_{max} trials had been executed during minimization of several functions of this particular test class. In these cases, the value of T_{max} equal to 1 000 000 was used in calculations of the average value in (4.26), providing in this way a lower estimate of the average. As can be seen from Tables 4.5 and 4.6, the methods based on the efficient partition strategy outperform DIRECT and DIRECT*l* also on Criterion C3.

Finally, the results of the comparison between the MULT*L* and MULT*K* algorithms and their two competitors in terms of Criterion C4 are reported in Table 4.7. This table shows how often the MULT*L* and MULT*K* methods were able to minimize each of 100 functions of a class with a smaller number of trials with respect to DIRECT or DIRECT*l*. The notation 'p:q' means that among 100 functions of a particular test class there are p functions for which DIRECT (or DIRECT*l*) spent fewer function trials than MULT*L* or MULT*K* and q functions for which MULT*L* or MULT*K* generated fewer trial points with respect to DIRECT (or DIRECT*l*) (p and q are from (4.27) and (4.28), respectively). For example, let us compare the MULT*L* method with DIRECT*l* on the simple GKLS two-dimensional class (see Table 4.7, the cell '52 : 47' in the first line). We can see that DIRECT*l* was better (was worse)

Table 4.2 Number of trial points generated by the DIRECT and DIRECT*l* algorithms and the MULT*L* and the MULT*K* methods during minimization of GKLS test functions (Criterion C1)

N	$\mathcal{D}$	50%						100%		
		DIRECT	DIRECT*l*	MULT*L*	MULT*K*	DIRECT	DIRECT*l*	MULT*L*	MULT*K*	
2	Simple	111	152	166	59	1159	2318	403	335	
2	Hard	1062	1328	613	182	3201	3414	1809	1075	
3	Simple	386	591	615	362	12507	13309	2506	2043	
3	Hard	1749	1967	1743	416	$> 10^6$ (4)	29233	6006	2352	
4	Simple	4805	7194	4098	2574	$> 10^6$ (4)	118744	14520	16976	
4	Hard	16114	33147	15064	3773	$> 10^6$ (7)	287857	42649	20866	
5	Simple	1660	9246	3854	1757	$> 10^6$ (1)	178217	33533	16300	
5	Hard	55092	126304	24616	13662	$> 10^6$ (16)	$> 10^6$ (4)	93745	88459	

Table 4.3 Number of hyperintervals generated by the DIRECT and DIRECT*l* algorithms and the MULT*L* and the MULT*K* methods during minimization of GKLS test functions (Criterion C2)

N	$\mathcal{D}$	50%				100%			
		DIRECT	DIRECT*l*	MULT*L*	MULT*K*	DIRECT	DIRECT*l*	MULT*L*	MULT*K*
2	Simple	111	152	269	185	1159	2318	685	1137
2	Hard	1062	1328	1075	607	3201	3414	3307	3993
3	Simple	386	591	1545	1867	12507	13309	6815	12149
3	Hard	1749	1967	5005	2061	$> 10^6$	29233	17555	14357
4	Simple	4805	7194	15145	21635	$> 10^6$	118744	73037	186295
4	Hard	16114	33147	68111	33173	$> 10^6$	287857	211973	223263
5	Simple	1660	9246	21377	19823	$> 10^6$	178217	206323	255059
5	Hard	55092	126304	177927	169413	$> 10^6$	$> 10^6$	735945	1592969

Table 4.4   Improvements obtained by the MULTL and MULTK in terms of Criterion C1

N	$\mathscr{D}$	$\dfrac{\text{DIRECT}}{\text{MULT}L}$	$\dfrac{\text{DIRECT}l}{\text{MULT}L}$	$\dfrac{\text{DIRECT}}{\text{MULT}K}$	$\dfrac{\text{DIRECT}l}{\text{MULT}K}$
2	Simple	2.88	5.75	3.46	6.92
2	Hard	1.77	1.89	2.98	3.18
3	Simple	4.99	5.31	6.12	6.51
3	Hard	>166.50	4.87	>425.17	12.43
4	Simple	>68.87	8.18	>58.91	6.99
4	Hard	>23.45	6.75	>47.92	13.80
5	Simple	>29.82	5.31	>61.35	10.93
5	Hard	>10.67	>10.67	>11.30	>11.30

Table 4.5   Average number of trial points generated by the DIRECT and DIRECTl algorithms and the MULTL method during minimization of GKLS test functions (Criterion C3)

N	$\mathscr{D}$	DIRECT	DIRECTl	MULTL	Improvement	
					$\dfrac{\text{DIRECT}}{\text{MULT}L}$	$\dfrac{\text{DIRECT}l}{\text{MULT}L}$
2	Simple	198.89	292.79	176.25	1.13	1.66
2	Hard	1063.78	1267.07	675.74	1.57	1.88
3	Simple	1117.70	1785.73	735.76	1.52	2.43
3	Hard	>42322.65	4858.93	2006.82	>21.09	2.42
4	Simple	>47282.89	18983.55	5014.13	>9.43	3.79
4	Hard	>95708.25	68754.02	16473.02	>5.81	4.17
5	Simple	>16057.46	16758.44	5129.85	>3.13	3.27
5	Hard	>217215.58	>269064.35	30471.83	>7.13	>8.83

than the new method on $p = 52$ ($q = 47$) functions of this class, and one problem was solved by the two methods with the same number of trials.

It can be seen that DIRECT and DIRECTl behave better than MULTL with respect to Criterion C4 when simple functions are minimized and the situation changes when hard classes are minimized. For example, for the hard GKLS two-dimensional class and DIRECTl we have '23 : 77' instead of '52 : 47' for the simple class (see the second and the first lines of Table 4.7, respectively). If a more difficult test class is taken, the MULTL method outperforms its two competitors (see hard classes of the dimensions $N = 2$, 4, and 5 in Table 4.7). For the three-dimensional classes DIRECT and DIRECTl were better than the MULTL method on this criterion (see Table 4.7). This happened because the hard three-dimensional class (although being more difficult than the simple one because the number q has increased in all the cases) continues to be too simple, especially, with respect to the increase of the difficulty manifested by other hard classes. Thus, since the new method is oriented on solving hard multidimensional multiextremal problems, the more hard objective functions are presented in a test class, the more pronounced is the advantage of the proposed

Table 4.6 Average number of trial points generated by the DIRECT and DIRECT*l* algorithms and the MULT*K* method during minimization of GKLS test functions (Criterion C3)

N	$\mathscr{D}$	DIRECT	DIRECT*l*	MULT*L*	Improvement	
					DIRECT MULT*L*	DIRECT*l* MULT*L*
2	Simple	198.89	292.79	97.22	2.06	3.01
2	Hard	1063.78	1267.07	192.00	5.54	6.60
3	Simple	1117.70	1785.73	491.28	2.28	3.63
3	Hard	>42322.65	4858.93	618.32	>68.45	7.86
4	Simple	>47282.89	18983.55	3675.84	>12.87	5.16
4	Hard	>95708.25	68754.02	5524.77	>17.32	12.44
5	Simple	>16057.46	16758.44	3759.05	>4.27	4.46
5	Hard	>217215.58	>269064.35	22189.47	>9.79	>12.13

Table 4.7 Comparison between the MULT*L* and the MULT*K* methods and the DIRECT and the DIRECT*l* methods in terms of Criterion C4

N	$\mathscr{D}$	DIRECT:MULT*L*	DIRECT*l*:MULT*L*	DIRECT:MULT*K*	DIRECT*l*:MULT*K*
2	Simple	61:39	52:47	28:72	21:79
2	Hard	36:64	23:77	15:85	16:84
3	Simple	66:34	54:46	36:64	30:70
3	Hard	58:42	51:49	19:81	17:83
4	Simple	51:49	37:63	39:61	25:75
4	Hard	47:53	42:58	14:86	16:84
5	Simple	66:34	26:74	55:45	17:83
5	Hard	34:66	27:73	26:74	20:80

algorithm. In its turn, the MULT*K* was always better than the competitors in terms of Criterion C4.

As demonstrated by the results of these extensive numerical experiments executed on 800 test functions, the methods based on the non-redundant diagonal partition strategy manifest quite a good performance. The usage of the gradient information together with the efficient partitioning strategy allows one to obtain a further serious acceleration in comparison with the DIRECT-based methods on the studied classes of test problems.

4.5 A Case Study: Fitting a Sum of Dumped Sinusoids to a Series of Observations

To illustrate the applicability of methods based on the non-redundant partition strategy, let us consider the following general nonlinear regression problem. Assume that we have a series of real-valued observations $y_1, \ldots, y_T$ such that

$$y_t = \eta(\theta, t) + \varepsilon_t, \quad 1 \le t \le T,$$

where θ is an N-dimensional vector of unknown parameters, $\eta(\theta, t)$ is a function nonlinear in θ and $\varepsilon_1, \ldots, \varepsilon_T$ is a series of noise terms (often assumed independently and identically distributed random variables, with zero mean and variance σ^2). The non-uniform sampling case $t = t_l, 1 \le l \le T$, can be studied similarly. Let $\Theta \subset \mathbb{R}^N$ be a parameter space so that $\theta \in \Theta$. Parameter estimation in the general nonlinear regression model can be reduced to solving the minimization problem

$$\min_{\theta \in \Theta} F(\theta), \quad F(\theta) = \sum_{t=1}^{T} (y_t - \eta(\theta, t))^2, \quad \Theta \subset \mathbb{R}^N, \tag{4.29}$$

with the estimator θ^* defined as

$$\theta^* = \arg \min_{\theta \in \Theta} F(\theta).$$

This problem can be often stated as a box-constrained global optimization problem, i.e., there exists a hyperinterval

$$\Theta = [\theta^-, \theta^+] = \{\theta \in \mathbb{R}^N : \theta^-(j) \le \theta(j) \le \theta^+(j), \ j = 1, \ldots, N\}. \tag{4.30}$$

We consider the case where the function $\eta(\theta, t)$ has the form

$$\eta(\theta, t) = \sum_{i=1}^{q} a_i \exp(d_i t) \sin(2\pi \omega_i t + \phi_i), 1 \le t \le T. \tag{4.31}$$

Here, q is a given integer, $\theta = (a, d, \omega, \phi)$ with $a = (a_1, \ldots, a_q), d = (d_1, \ldots, d_q),$ $\omega = (\omega_1, \ldots, \omega_q)$ and $\phi = (\phi_1, \ldots, \phi_q)$. Denote the true (to be estimated) vector of parameters by $\theta^{(0)} = (a^{(0)}, d^{(0)}, \omega^{(0)}, \phi^{(0)})$ ($\theta^{(0)}$ coincides with θ^* in the case of noise-free observations). The range for parameters a_i and d_i is $(-\infty, \infty)$, whilst the ranges for ω_i and ϕ_i are $[0, 1)$ and $[0, \pi/2)$, respectively.

The model (4.31) is a very well-known model which in the signal processing literature is called the 'sums of damped sinusoids' (see, e.g., [31, 42, 52, 110, 190] for some typical techniques used in the signal processing field). This model is also associated with the so-called Hankel structured low-rank approximation problem which is described as follows (see, e.g., [121, 124]). Let $X = (x_{i,j})$ be an $L \times K$ matrix such that $x_{i,j} = y_{i+j-1}$ and $T + 1 = L + K$. Then, the matrix X is of Hankel structure. The Hankel structured low-rank approximation problem is that of finding another Hankel matrix 'close' to X (the most common instance of the problem uses the Frobenius norm) which is of some pre-specified rank $r < \min(L, K)$. Full details describing the formal connection are available, e.g., in [122, 123, 125].

In this section, we show that the optimization problem (4.29) becomes challenging if the general model (4.31) is considered. The following two particular cases of the model (4.31) will be also studied:

$$\eta(a, \omega, \phi; t) = \sum_{i=1}^{q} a_i \sin(2\pi \omega_i t + \phi_i), \quad 1 \leq t \leq T, \tag{4.32}$$

with $d = (0, 0, \ldots, 0)^T$, and

$$\eta(a, d, \phi; t) = \sum_{i=1}^{q} a_i \exp(d_i t) \sin(2\pi \omega_i^{(0)} t + \phi_i), \ 1 \leq t \leq T, \tag{4.33}$$

where it is assumed that the vector ω is known: $\omega = \omega^{(0)}$. The following sets of parameters are studied:

(i) $\theta = (a, d, \omega, \phi)$, the general case.
(ii) $\theta = (a, \omega, \phi)$, so that it is assumed that the vector d is known: $d = d^{(0)}$. In many applications, $d = (0, 0, \ldots, 0)^T$.
(iii) $\theta = (a, d, \phi)$, so that it is assumed that the vector ω is known: $\omega = \omega^{(0)}$.

The problem (i) is a classical problem important in spectral analysis (see, e.g., [126, 247]) and many applications, such as magnetic resonance spectroscopy, radioastronomy, antenna theory, prospecting seismology, and so on (see, e.g., [56, 182, 183, 247]). The problem (ii) with $d = 0$ can be considered as a simple extension of the Fourier expansion and hence can be applied in many different fields (see, e.g., [31, 110]). As extension of the Fourier model, it would be especially valuable when the frequencies ω_i are not expected to necessarily be $1/k$ (this includes the case of the so-called quasi-periodic signals). The problem (iii) naturally appears in models where the frequencies are known (see, e.g., [247]). A typical example would be provided by a monthly economic activity time series where the most dominant frequency is 12 supplemented with fractions $12/k$ with $k = 2, 3, 4, 5$, and 6.

In the literature, discussions on the behavior of the objective function (4.29) can be found, e.g., in [123]. In that paper, the fact that the objective function $F(\theta)$ is multiextremal has been observed; the function $F(\theta)$ was decomposed into three different components and it was numerically demonstrated that the part of the objective function with the observation noise removed dominates the shape of the objective function. The optimization problem (4.29) is very difficult (even in the case $q = 1$ in (4.31)–(4.33)) with the objective function possessing many local minima. Although the objective function is Lipschitz-continuous (see, e.g., [125, 291, 323, 348]), it has very high Lipschitz constants which increase with T, the number of observations. Additionally, increasing T leads to more erratic objective functions with a higher number of local minima. Adding noise to the observed data increases the complexity of the objective function (see, e.g., [40, 351]) and moves the global minimizer away from the vector of true parameters. Thus, efficient global optimization techniques should be used to tackle the stated problem.

4.5.1 Examples Illustrating the Complexity of the Problem

Let us provide a number of examples (see, e.g., [120, 296]) to illustrate the complexity of the problem (4.29) with the function given by (4.31)–(4.33). Figure 4.6 contains plots of the objective function for the following cases:

(a) Sine with unknown frequency only: we take $q = 1$ (the simplest case) and $T = 10$ in (4.31) and consider a particular objective function of the form

$$F(\omega_1) = \sum_{t=1}^{T} (y_t - \sin(2\pi\omega_1 t))^2 , \qquad (4.34)$$

where $\omega^{(0)} = \omega_1^{(0)} = 0.4$ (see Fig. 4.6a). This one-dimensional function $F(\omega_1)$ is periodic with period 1 and the minimum value $F^* = 0$ (noise-free observations are considered for simplicity) is attained at the points $\omega_1^* = \omega_1^{(0)} + p$ ($p = 0, \pm1, \pm2, \ldots$). The feasible domain for ω_1 can therefore be chosen as $[0, 1) \subset \mathbb{R}^1$; in this interval, the function $F(\omega_1)$ has one global minimizer at $\omega_1^* = \omega_1^{(0)}$ and many local minimizers.

(b) We repeat the above but take a higher $T = 100$ (see Fig. 4.6b).

In both the cases (a) and (b), it can be easily seen that the objective function $F(\omega_1)$ is highly multiextremal and very irregular (for the parameter settings in these examples, the number of local minimizers increases linearly in T). For $T = 10$, the Lipschitz constant of $F(\omega_1)$ (estimated over 10^{-7}-grid) is approximately equal to 432.0. For $T = 100$, it becomes equal to $\simeq 28690.8$. The global minimizer has a very narrow attraction region and it moves away from the true parameters vector $\theta^{(0)}$ in (4.31) when the observations y_t are measured with noise. Therefore, already in these relatively simple cases ($q = 1$ in (4.31)) a particular care should be taken in the choice of global optimization algorithms to find this global minimizer.

(c) Two sines with unknown frequencies only: we take $q = 2$ in (4.31), $T = 10$ noise-free observations in (4.29) and consider the following objective function over the admissible domain $\omega = (\omega_1, \omega_2) \in [0, 1) \times [0, 1)$:

$$F(\omega_1, \omega_2) = \sum_{t=1}^{T} (y_t - \sin(2\pi\omega_1 t) - \sin(2\pi\omega_2 t))^2 , \qquad (4.35)$$

where $\omega_1^{(0)} = 0.3$, $\omega_2^{(0)} = 0.4$ (see Fig. 4.6c).

The two-dimensional objective function $F(\omega_1, \omega_2)$ is again highly multiextremal (note its symmetry with respect to permutation of the parameters). Although two global minimizers are well separated (in terms of the objective function values too) from the multitude of local minimizers, the effect of having two close frequencies can be seen in Fig. 4.6c.

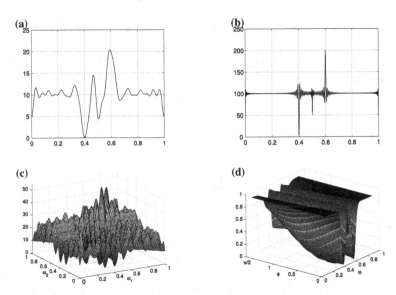

Fig. 4.6 Benchmark objective functions in our case study: **a** $F(\omega_1)$ from (4.34) with $T = 10$; **b** $F(\omega_1)$ from (4.34) with $T = 100$; **c** $F(\omega_1, \omega_2)$ from (4.35) with $T = 10$; **d** $F(\omega_1, \omega_2)$ from (4.35) with $T = 10$

(d) Sine with unknown frequency, phase, and amplitude: we consider again the optimization problem (4.29) defined by the model (4.31) with $q = 1$. As (4.31) is quadratic in the parameter a, it is possible (and sometimes can be useful) to obtain an explicit estimator for a based on the remaining parameters d, ω, and ϕ (as shown in the following), thus defining the three-dimensional objective function $f(d, \omega, \phi)$ for $d \in [-2, 2] \subset \mathbb{R}^1$, $\omega \in (0, 1)$, and $\phi \in (0, \pi/2)$. In Fig. 4.6d, its cross-section (ω, ϕ) with $d = d^{(0)} = -0.2$ in the case of $T = 10$ noise-free observations is plotted, with true values $a^{(0)} = 1.0$, $d^{(0)} = -0.2$, $\omega^{(0)} = 0.4$, and $\phi^{(0)} = 0.3$. A highly multiextremal behaviour of this three-dimensional objective function can be again observed, especially with respect to the frequency ω.

The described objective functions are Lipschitz-continuous with high Lipschitz constants, essentially multiextremal and derivable. Their evaluation is often associated (especially when a high number T of noisy observations are given) with performing computationally expensive numerical experiments. Moreover, noisy observations in (4.29) furthermore increase the problem complexity by both shifting the global minimizer away from the true parameters values and making unknown the desired global minimum value F^* (which is equal to $F^* = 0$ in the case of noise-free observations).

4.5.2   Derivatives and Simplifications of the Benchmark Objective Functions

Let us consider the optimization problem (4.29) defined by the model (4.31) with different unknown parameters, as described above.

If $q = 1$ (the simplest case of one sine function in (4.31)) and only one frequency ω_1 is unknown (the cases (a) and (b)), the first derivative of the objective function $F(\omega_1)$ from (4.34) can be explicitly written over $\omega_1 \in [0, 1)$ as

$$F'(\omega_1) = -4\pi \sum_{t=1}^{T} t \cos(2\pi\omega_1 t)(y_t - \sin(2\pi\omega_1 t)). \tag{4.36}$$

If two sine functions ($q = 2$ in (4.31)) with two unknown frequencies ω_1 and ω_2 are considered (the case (c) with $F(\omega_1, \omega_2)$ from (4.35)), then

$$\frac{\partial F}{\partial \omega_1} = -4\pi \sum_{t=1}^{T} t \cos(2\pi\omega_1 t)(y_t - \sin(2\pi\omega_1 t) - \sin(2\pi\omega_2 t)),$$

$$\frac{\partial F}{\partial \omega_2} = -4\pi \sum_{t=1}^{T} t \cos(2\pi\omega_2 t)(y_t - \sin(2\pi\omega_1 t) - \sin(2\pi\omega_2 t)).$$

$$\tag{4.37}$$

Let us finally consider the case (d), with $q = 1$ in (4.31) but unknown frequency, phase, and amplitude parameters in the dumped sinusoid (4.31). For brevity, let us take $x_t = \exp(dt) \sin(2\pi\omega t + \phi)$. As a result, equation (4.29) may be written as

$$F(a, d, \omega, \phi) = \sum_{t=1}^{T} (y_t - a x_t)^2. \tag{4.38}$$

Then, since

$$\frac{\partial F(a, d, \omega, \phi)}{\partial a} = -2 \sum_{t=1}^{T} (y_t - a x_t) x_t,$$

we may obtain ($x_t \neq 0$ for all $t = 1, \ldots, T$) an explicit estimator for a, which we denote $\hat{a}$. This estimator is a function of the remaining parameters d, ω, and ϕ:

$$\hat{a} = \frac{\sum_{t=1}^{T} y_t x_t}{\sum_{t=1}^{T} x_t^2}.$$

By substituting $\hat{a}$ into (4.38) (such a transformation can be used to decrease the number of independent parameters), we obtain a new objective function, which we denote $f(d, \omega, \phi)$:

$$f(d, \omega, \phi) = \sum_{t=1}^{T} \left(y_t - x_t \frac{\sum_{k=1}^{T} y_k x_k}{\sum_{k=1}^{T} x_k^2} \right)^2 . \tag{4.39}$$

It is possible to compute the derivatives of the objective function with respect to each of the unknown parameters, although they cannot be always written in a neat form. Here, we state only the first derivatives of $f(d, \omega, \phi)$ with respect to each of the unknown parameters:

$$\frac{\partial f}{\partial d} = -2 \sum_{t=1}^{T} \left\{ \left[y_t - \frac{x_t \sum_{k=1}^{T} y_k x_k}{\sum_{k=1}^{T} x_k^2} \right] \times \right.$$
$$\left. \left[\frac{x_t \sum_{k=1}^{T} k y_k x_k}{\sum_{k=1}^{T} x_k^2} + \frac{x_t \sum_{k=1}^{T} y_k x_k}{\left(\sum_{k=1}^{T} x_k^2 \right)^2} \left(t \sum_{k=1}^{T} x_k^2 - 2 \sum_{k=1}^{T} k x_k^2 \right) \right] \right\} . \tag{4.40}$$

If we take

$$c_1^{(t)} = t \exp(dt) \cos(2\pi \omega t + \phi) \sum_{k=1}^{T} x_k^2 - 2x_t \sum_{k=1}^{T} k x_k \exp(dk) \cos(2\pi \omega k + \phi),$$

then, it follows

$$\frac{\partial f}{\partial \omega} = -4\pi \sum_{t=1}^{T} \left\{ \left[y_t - \frac{x_t \sum_{k=1}^{T} y_k x_k}{\sum_{k=1}^{T} x_k^2} \right] \times \right.$$
$$\left. \left[\frac{x_t \sum_{k=1}^{T} y_k k \exp(dk) \cos(2\pi \omega k + \phi)}{\left(\sum_{k=1}^{T} x_k^2 \right)^2} + \frac{\sum_{k=1}^{T} y_k x_k}{\left(\sum_{k=1}^{T} x_k^2 \right)^2} c_1^{(t)} \right] \right\} . \tag{4.41}$$

If we take

$$c_2^{(t)} = \exp(dt) \cos(2\pi \omega t + \phi) \sum_{k=1}^{T} x_k^2 - 2x_t \sum_{k=1}^{T} x_k \exp(dk) \cos(2\pi \omega k + \phi),$$

then, it follows

$$\frac{\partial f}{\partial \phi} = -2 \sum_{t=1}^{T} \left\{ \left[y_t - \frac{x_t \sum_{k=1}^{T} y_k x_k}{\sum_{k=1}^{T} x_k^2} \right] \times \right.$$
$$\left. \left[\frac{x_t \sum_{k=1}^{T} y_k \exp(dk) \cos(2\pi \omega k + \phi)}{\left(\sum_{k=1}^{T} x_k^2 \right)^2} + \frac{\sum_{k=1}^{T} y_k x_k}{\left(\sum_{k=1}^{T} x_k^2 \right)^2} c_2^{(t)} \right] \right\} .$$
$$\tag{4.42}$$

Table 4.8 Solutions to
benchmark problems obtained
by the MULTK method

$F(\theta)$	MULTK
(a)	$\omega_1^* \approx 0.400000$
	# trials = 43
(b)	$\omega_1^* \approx 0.400000$
	# trials = 170
(c)	$\omega_1^* \approx 0.299040$
	$\omega_2^* \approx 0.400549$
	# trials = 204
(d)	$d^* \approx -0.222222$
	$\omega^* \approx 0.395062$
	$\phi^* \approx 0.310123$
	# trials = 1449

It can be shown that all the derivatives (4.36), (4.37), and (4.40)–(4.42) are Lipschitz-continuous functions (generally, with very high but unknown Lipschitz constants) over suitably defined domains of the parameters.

4.5.3   Numerical Examples and Simulation Study

Let us now report some results (see [120]) of numerical experiments performed with the method MULTK (described in Sect. 4.3) on the benchmark functions presented above to illustrate its performance on the applied problems (4.29)–(4.33).

Closed admissible domains Θ from (4.30) were considered for all problems. Since in the case of noise-free benchmark problems the solutions are known ($\theta^* = \theta^0$ in (4.29)), a particular test problem was considered to be solved by the method if it generated a trial point θ' in an ε-neighborhood of θ^* (precisely, θ' should be inside a hyperinterval with a vertex θ^* having the volume smaller than the volume of the initial hyperinterval Θ from (4.30) multiplied by ε; see, e.g., [179] for more details). In the case of real-life noisy problems, this method can be stopped (as often adopted in practical algorithms) when a prescribed number of trials is reached: the current best value is thus taken as a solution to the problem.

In the experiments performed, the accuracy ε was taken at least equal to 10^{-6}, that can be often considered an acceptable accuracy in practice. The balancing parameter (cf. with (4.23)) equal to 10^{-4} was used in the MULTL and MULTK methods (this parameter prevents the algorithms from subdividing too small hyperintervals; see, e.g., [102, 233]).

Results of numerical experiments are reported in Table 4.8 where for all the cases both the found solution and the number of generated trial points are given. It can be seen from this Table that the found parameters were quite accurate and were determined by the considered method within a very limited budget of trials.

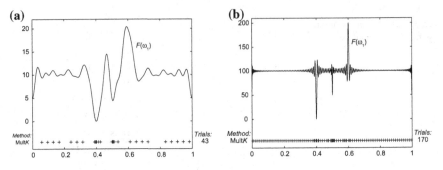

Fig. 4.7 Distribution of trial points generated MULTK method by when solving one-dimensional problems: **a** $F(\omega_1)$ from (4.34) with $T = 10$; **b** $F(\omega_1)$ from (4.34) with $T = 100$

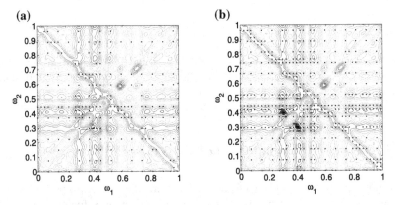

Fig. 4.8 Distribution of trial points generated by MULTK when solving two-dimensional problem (c) $F(\omega_1, \omega_2)$ from (4.35): **a** accuracy $\varepsilon = 10^{-6}$, number of the first successful trial is 204; **b** accuracy $\varepsilon = 10^{-7}$, number of the first successful trial is 540

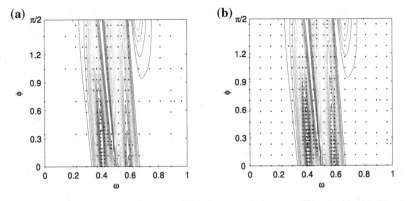

Fig. 4.9 Distribution of trial points generated by MULTK when solving three-dimensional problem (d) $f(d, \omega, \phi)$ from (4.39): **a** accuracy $\varepsilon = 10^{-6}$, number of the first successful trial is 1449; **b** accuracy $\varepsilon = 10^{-7}$, number of the first successful trial is 4015

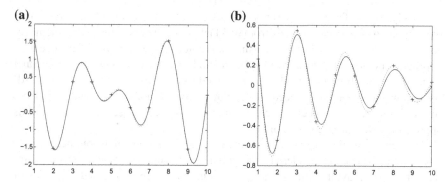

Fig. 4.10 Plots of observations y_t (signed by $+$), true signals (dotted lines), and the reconstructed signals (solid lines) estimated by the MULTK method for benchmark problems: **a** true signal $\eta(\omega_1^{(0)}, \omega_2^{(0)})$ from (4.35), $\omega_1^{(0)} = 0.3$, $\omega_2^{(0)} = 0.4$, and the estimated reconstructed signal $\eta(\omega_1^*, \omega_2^*)$ with $\omega_1^* \approx 0.300412$, $\omega_2^* \approx 0.399177$; **b** true signal $\eta(d^{(0)}, \omega^{(0)}, \phi^{(0)})$ from (4.39), $d^{(0)} = -0.2$, $\omega^{(0)} = 0.3$, $\phi^{(0)} = 0.4$, and the estimated reconstructed signal $\eta(d^*, \omega^*, \phi^*)$ with $d^* \approx -0.222222$, $\omega^* \approx 0.395062$, $\phi^* \approx 0.290741$

Trial points generated by the MULTK method when solving one-dimensional and multidimensional benchmark problems are reported in Figs. 4.7, 4.8 and 4.9, respectively, together with the objective function contours. Convergence of the sequences of trial points generated by the method to the global minimizers can be thus evidenced and differences between two values of the accuracy ε can be observed.

For example, the situations where the MULTK method generated the first trial point in an ε-neighborhood of the known global minimizer of the two-dimensional function $F(\omega_1, \omega_2)$ from (4.35) are illustrated in Fig. 4.8a. Recall that such an artificial stopping rule (applicable for benchmark problems) cannot be used in practical real-life problems. In order to evaluate the real behavior of the methods, more trials are needed. For the MULTK, this can be done either by allowing the method to run until achieving some computational budget, or (equivalently) by decreasing the accuracy ε. Note that trial points were concentrated around global minimizers $(0.3, 0.4)$ and $(0.4, 0.3)$ of this two-dimensional function $F(\omega_1, \omega_2)$ which is symmetric in this case with respect to permutation of its arguments ω_1 and ω_2. Closer were the frequencies $\omega_1^{(0)}$ and $\omega_2^{(0)}$, less trials were needed to locate the best parameters. Global optimization algorithms that are able to make use of the symmetries in the objective function (e.g., simplicial partitioning methods; see [233, 235, 348]) could be more advantageous for optimizing such particular functions.

Similar conclusions with regard to the methods' performance can be also given in the case of the three-dimensional benchmark function $f(d, \omega, \phi)$ from (4.39). Its two-dimensional cross-section (ω, ϕ) for the true parameter $d^{(0)} = -0.2$ with trial points generated (and projected to the taken cross-section) by the method at its different stages is reported in Fig. 4.9.

Finally, Fig. 4.10a and b report realizations of observations y_t and the corresponding true and reconstructed signals from (4.31) in the case of benchmark problems (c) and (d), respectively (the case of benchmark problems (a) and (b) is trivial since their optimal value ω_1^* was determined by the methods exactly up to the seventh decimal point). It can be observed how the reconstructed signal fitted closely the true one. All the optimal parameters produced by the considered method can be (and should be) further improved by a specific local optimization procedure (see, e.g., [102]), thus approaching the true signals even better.

Thus, in this section, a classical parameter estimation problem in nonlinear regression models has been considered as a global optimization problem. Several examples of the objective functions stated to fit sums of dumped sinusoids to series of observations have been particularly examined to illustrate the complexity of this identification problem. For its study, the usage of deterministic global optimization techniques has been proposed, since they can often provide solutions to these difficult problems together with some guaranteed gaps. Particularly, one of the promising Lipschitz-based methods has been successfully applied to determine the solutions of the analyzed problems.

To conclude this small book we would like to stress that the real life is plenty of global optimization problems. We wish the reader to successfully find all the global optima and hope that the diagonal approach briefly described in the present book can be useful for reaching this goal.

References

1. Addis, B., Locatelli, M.: A new class of test functions for global optimization. J. Glob. Optim. **38**(3), 479–501 (2007)
2. Adjiman, C.S., Dallwig, S., Floudas, C.A., Neumaier, A.: A global optimization method, αBB, for general twice-differentiable constrained NLPs - I. Theor. Adv. Comput. Chem. Eng. **22**(9), 1137–1158 (1998)
3. Afraimovich, L.G., Katerov, A.S., Prilutskii, M.K.: Multi-index transportation problems with 1-nested structure. Autom. Remote Control **77**(11), 1894–1913 (2016)
4. Aguiar e Oliveira Jr., H., Ingber, L., Petraglia, A., Rembold Petraglia, M., Augusta Soares Machado, M. (eds.): Stochastic Global Optimization and Its Applications with Fuzzy Adaptive Simulated Annealing. Intelligent Systems Reference Library, vol. 35. Springer (2012)
5. Andramonov, M.Y., Rubinov, A.M., Glover, B.M.: Cutting angle methods in global optimization. Appl. Math. Lett. **12**(3), 95–100 (1999)
6. Androulakis, I.P., Maranas, C.D., Floudas, C.A.: αBB: a global optimization method for general constrained nonconvex problems. J. Glob. Optim. **7**(4), 337–363 (1995)
7. Archetti, F., Schoen, F.: A survey on the global optimization problem: general theory and computational approaches. Ann. Oper. Res. **1**(2), 87–110 (1984)
8. Astorino, A., Frangioni, A., Gaudioso, M., Gorgone, E.: Piecewise-quadratic approximation in convex numerical optimization. SIAM J. Optim. **21**(4), 1418–1438 (2011)
9. Astorino, A., Fuduli, A., Gaudioso, M.: DC models for spherical separation. J. Glob. Optim. **48**(4), 657–669 (2010)
10. Audet, C., Hansen, P., Savard, G. (eds.): Essays and surveys in global optimization. GERAD 25th Anniversary. Springer, New York (2005)
11. Bagirov, A.M., Rubinov, A.M., Zhang, J.: Local optimization method with global multidimensional search. J. Glob. Optim. **32**(2), 161–179 (2005)
12. Baker, C.A., Watson, L.T., Grossman, B., Mason, W.H., Haftka, R.T.: Parallel global aircraft configuration design space exploration. In: Paprzycki, M., Tarricone, L., Yang, L.T. (eds.) Practical parallel computing. Special Issue of the International Journal of Computer Research, vol. 10(4), pp. 79–96. Nova Science Publishers, Hauppauge, NY (2001)
13. Baritompa, W.: Customizing methods for global optimization–a geometric viewpoint. J. Glob. Optim. **3**(2), 193–212 (1993)
14. Baritompa, W.: Accelerations for a variety of global optimization methods. J. Glob. Optim. **4**(1), 37–45 (1994)
15. Baritompa, W., Cutler, A.: Accelerations for global optimization covering methods using second derivatives. J. Glob. Optim. **4**(3), 329–341 (1994)
16. Barkalov, K., Polovinkin, A., Meyerov, I., Sidorov, S., Zolotykh, N.: SVM regression parameters optimization using parallel global search algorithm. In: Malyshkin, V. (ed.) Parallel Computing Technologies (PaCT 2013). LNCS, vol. 7979, pp. 154–166. Springer, New York (2013)

17. Barkalov, K.A., Gergel, V.P.: Parallel global optimization on GPU. J. Glob. Optim. **66**(1), 3–20 (2016)

18. Barkalov, K.A., Strongin, R.G.: A global optimization technique with an adaptive order of checking for constraints. Comput. Math. Math. Phys. **42**(9), 1289–1300 (2002)

19. Barkalov, K.A., Sysoyev, A.V., Lebedev, I.G., Sovrasov, V.V.: Solving GENOPT problems with the use of ExaMin solver. In: Festa, P., Sellmann, M., Vanschoren, J. (eds.) Learning and Intelligent Optimization (LION 2016). LNCS, vol. 10079, pp. 283–295. Springer, New York (2016)

20. Bartholomew-Biggs, M.C., Parkhurst, S.C., Wilson, S.P.: Using DIRECT to solve an aircraft routing problem. Comput. Optim. Appl. **21**(3), 311–323 (2002)

21. Bartholomew-Biggs, M.C., Ulanowski, Z.J., Zakovic, S.: Using global optimization for a microparticle identification problem with noisy data. J. Glob. Optim. **32**(3), 325–347 (2005)

22. Basso, P.: Iterative methods for the localization of the global maximum. SIAM J. Numer. Anal. **19**(4), 781–792 (1982)

23. Batishchev, D.I.: Search Methods of Optimal Design. Sovetskoe Radio, Moscow (1975). In Russian

24. Battiti, R., Sergeyev, Y.D., Brunato, M., Kvasov, D.E.: GENOPT-2016: Design of a GENeralization-based challenge in global OPTimization. In: Sergeyev, Y.D., et al. (ed.) Numerical Computations: Theory and Algorithms (NUMTA-2016). AIP Conference Proceedings, vol. 1776, p. 060005. AIP Publishing, New York (2016)

25. Beliakov, G.: Cutting angle method–a tool for constrained global optimization. Optim. Methods Softw. **19**(2), 137–151 (2004)

26. Beliakov, G., Ferrer, A.: Bounded lower subdifferentiability optimization techniques: applications. J. Glob. Optim. **47**(2), 211–231 (2010)

27. Bertolazzi, P., Guerra, C., Liuzzi, G.: A global optimization algorithm for protein surface alignment. BMC Bioinf. **11**, 488–498 (2010)

28. Bertsekas, D.P.: Nonlinear Programming. Athena Scientific, Massachusetts (1999)

29. Betrò, B.: Bayesian methods in global optimization. J. Glob. Optim. **1**(1), 1–14 (1991)

30. Björkman, M., Holmström, K.: Global optimization of costly nonconvex functions using radial basis functions. Optim. Eng. **1**(4), 373–397 (2000)

31. Bloomfield, P. (ed.): Fourier Analysis of Time Series: An Introduction. Wiley, New York (2000)

32. Bomze, I.M., Csendes, T., Horst, R., Pardalos, P.M. (eds.): Developments in Global Optimization. Kluwer Academic Publishers, Dordrecht (1997)

33. Bravi, L., Piccialli, V., Sciandrone, M.: An optimization-based method for feature ranking in nonlinear regression problems. IEEE Trans. Neural Netw. Learn. Syst. **28**(4), 1005–1010 (2017)

34. Breiman, L., Cutler, A.: A deterministic algorithm for global optimization. Math. Program. **58**(1–3), 179–199 (1993)

35. Brunato, M., Battiti, R.: A telescopic binary learning machine for training neural networks. IEEE Trans. Neural Netw. Learn. Syst. **28**(3), 665–677 (2017)

36. Bulatov, V.P.: Methods of embedding-cutting off in problems of mathematical programming. J. Glob. Optim. **48**(1), 3–15 (2010)

37. Butz, A.R.: Space filling curves and mathematical programming. Inform. Control **12**(4), 314–330 (1968)

38. Calvin, J.M.: A lower bound on convergence rates of nonadaptive algorithms for univariate optimization with noise. J. Glob. Optim. **48**(1), 17–27 (2010)

39. Calvin, J.M., Žilinskas, A.: On the convergence of the P-algorithm for one-dimensional global optimization of smooth functions. J. Optim. Theory Appl. **102**(3), 479–495 (1999)

40. Calvin, J.M., Žilinskas, A.: One-dimensional global optimization for observations with noise. Comput. Math. Appl. **50**(1–2), 157–169 (2005)

41. Campana, E.F., Diez, M., Iemma, U., Liuzzi, G., Lucidi, S., Rinaldi, F., Serani, A.: Derivative-free global ship design optimization using global/local hybridization of the DIRECT algorithm. Optim. Eng. **17**(1), 127–156 (2016)

42. Carnì, D.L., Fedele, G.: Multi-sine fitting algorithm enhancement for sinusoidal signal characterization. Comput. Stand. Int. **34**(6), 535–540 (2012)
43. Carr, C.R., Howe, C.W.: Quantitative Decision Procedures in Management and Economic: Deterministic Theory and Applications. McGraw-Hill, New York (1964)
44. Carter, R.G., Gablonsky, J.M., Patrick, A., Kelley, C.T., Eslinger, O.J.: Algorithms for noisy problems in gas transmission pipeline optimization. Optim. Eng. **2**(2), 139–157 (2001)
45. Cartis, C., Fowkes, J.M., Gould, N.I.M.: Branching and bounding improvements for global optimization algorithms with Lipschitz continuity properties. J. Glob. Optim. **61**(3), 429–457 (2015)
46. Casado, L.G., García, I., Csendes, T.: A new multisection technique in interval methods for global optimization computing. Computing **65**(3), 263–269 (2000)
47. Casado, L.G., García, I., Sergeyev, Y.D.: Interval branch and bound algorithm for finding the first-zero-crossing-point in one-dimensional functions. Reliab. Comput. **6**(2), 179–191 (2000)
48. Casado, L.G., García, I., Sergeyev, Y.D.: Interval algorithms for finding the minimal root in a set of multiextremal non-differentiable one-dimensional functions. SIAM J. Sci. Comput. **24**(2), 359–376 (2002)
49. Clarke, F.H.: Optimization and Nonsmooth Analysis. Wiley, New York (1983). (Reprinted by SIAM Publications, 1990)
50. Clausen, J., Žilinskas, A.: Subdivision, sampling, and initialization strategies for simplical branch and bound in global optimization. Comput. Math. Appl. **44**(7), 943–955 (2002)
51. Cochran, J.J. (ed.): Wiley Encyclopedia of Operations Research and Management Science (8 Volumes). Wiley, New York (2011)
52. Coluccio, L., Eisinberg, A., Fedele, G.: A Prony-like method for non-uniform sampling. Signal Process. **87**(10), 2484–2490 (2007)
53. Conn, A.R., Scheinberg, K., Vicente, L.N.: Introduction to Derivative-Free Optimization. SIAM, Philadelphia (2009)
54. Corliss, G., Faure, C., Griewank, A., Hascoet, L., Naumann, U. (eds.): Automatic Differentiation of Algorithms: From Simulation to Optimization. Springer, New York (2002)
55. Cormen, T.H., Leiserson, C.E., Rivest, R.L., Stein, C.: Introduction to Algorithms, 2nd edn. MIT Press and McGraw-Hill, New York (2001)
56. Costanzo, S., Venneri, I., Di Massa, G., Borgia, A.: Benzocyclobutene as substrate material for planar millimeter-wave structures: dielectric characterization and application. J. Infrared Milli Terahz Waves **31**, 66–77 (2010)
57. Cox, S.E., Haftka, R.T., Baker, C.A., Grossman, B., Mason, W.H., Watson, L.T.: A comparison of global optimization methods for the design of a high-speed civil transport. J. Glob. Optim. **21**(4), 415–433 (2001)
58. Csallner, A.E., Csendes, T., Markót, M.C.: Multisection in interval branch-and-bound methods for global optimization - I. Theor. Results J. Glob. Optim. **16**(4), 371–392 (2000)
59. Csendes, T. (ed.): Developments in Reliable Computing. Kluwer Academic Publishers, Dordrecht (2000)
60. Csendes, T., Ratz, D.: Subdivision direction selection in interval methods for global optimization. SIAM J. Numer. Anal. **34**(3), 922–938 (1997)
61. Custódio, A.L.: Aguilar Madeira, J.F.: GLODS: global and local optimization using direct search. J. Glob. Optim. **62**(1), 1–28 (2015)
62. Custódio, A.L.: Aguilar Madeira, J.F., Vaz, A.I.F., Vicente, L.N.: Direct multisearch for multiobjective optimization. SIAM J. Optim. **21**(3), 1109–1140 (2011)
63. Danilin, Y.M.: Estimation of the efficiency of an absolute-minimum-finding algorithm. USSR Comput. Math. Math. Phys. **11**(4), 261–267 (1971)
64. Daponte, P., Grimaldi, D., Molinaro, A., Sergeyev, Y.D.: An algorithm for finding the zero-crossing of time signals with lipschitzean derivatives. Measurement **16**(1), 37–49 (1995)
65. Daponte, P., Grimaldi, D., Molinaro, A., Sergeyev, Y.D.: Fast detection of the first zero-crossing in a measurement signal set. Measurement **19**(1), 29–39 (1996)

66. Davies, G., Gillard, J.W., Zhigljavsky, A.: Comparative study of different penalty functions and algorithms in survey calibration. In: Pardalos, P.M., Zhigljavsky, A., Žilinskas, J. (eds.) Advances in Stochastic and Deterministic Global Optimization. Springer Optimization and Its Applications, vol. 107, pp. 87–127. Springer, Switzerland (2016)

67. De Cosmis, S., De Leone, R.: The use of grossone in mathematical programming and operations research. Appl. Math. Comput. **218**(16), 8029–8038 (2012)

68. Demyanov, V.F., Malozemov, V.N.: Introduction to Minimax. Wiley, New York (1974). (The 2nd English–language edition: Dover Publications, 1990)

69. Demyanov, V.F., Rubinov, A.M.: Quasidifferential Calculus. Optimization Software Inc., Publication Division, New York (1986)

70. Di Pillo, G., Giannessi, F. (eds.): Nonlinear Optimization and Applications. Plenum Press, New York (1996)

71. Di Pillo, G., Grippo, L.: An exact penalty function method with global convergence properties for nonlinear programming problems. Math. Program. **36**(1), 1–18 (1986)

72. Di Pillo, G., Liuzzi, G., Lucidi, S., Piccialli, V., Rinaldi, F.: Univariate global optimization with multiextremal non-differentiable constraints without penalty functions. Comput. Optim. Appl. **65**(2), 361–397 (2016)

73. Di Serafino, D., Gomez, S., Milano, L., Riccio, F., Toraldo, G.: A genetic algorithm for a global optimization problem arising in the detection of gravitational waves. J. Glob. Optim. **48**(1), 41–55 (2010)

74. Di Serafino, D., Liuzzi, G., Piccialli, V., Riccio, F., Toraldo, G.: A modified dividing rectangles algorithm for a problem in astrophysics. J. Optim. Theory Appl. **151**(1), 175–190 (2011)

75. Dixon, L.C.W.: Global optima without convexity. Technical Report, Numerical Optimization Centre, Hatfield Polytechnic, Hatfield, England (1978)

76. Dixon, L.C.W., Szegö, G.P. (eds.): Towards Global Optimization (volumes 1 and 2). North–Holland, Amsterdam (1975, 1978)

77. Dolan, E.D., Moré, J.J.: Benchmarking optimization software with performance profiles. Math. Progr. **91**, 201–213 (2002)

78. Dzemyda, G., Šaltenis, V., Žilinskas, A. (eds.): Stochastic and Global Optimization. Kluwer Academic Publishers, Dordrecht (2002)

79. Elsakov, S.M., Shiryaev, V.I.: Homogeneous algorithms for multiextremal optimization. Comp. Math. Math. Phys. **50**(10), 1642–1654 (2010)

80. Ermoliev, Y.M., Wets, R.J.B. (eds.): Numerical Techniques for Stochastic Optimization Problems. Springer, Berlin (1988)

81. Evtushenko, Y.G.: Numerical methods for finding global extrema (case of a non-uniform mesh). USSR Comput. Math. Math. Phys. **11**(6), 38–54 (1971)

82. Evtushenko, Y.G.: Numerical Optimization Techniques. Translations Series in Mathematics and Engineering. Springer, Berlin (1985)

83. Evtushenko, Y.G., Lurie, S.A., Posypkin, M.A.: New optimization problems arising in modelling 2D-crystal lattices. In: Sergeyev, Y.D., et al. (ed.) Numerical Computations: Theory and Algorithms (NUMTA-2016). In: AIP Conference Proceedings, vol. 1776, p. 060007. AIP Publishing, New York (2016)

84. Evtushenko, Y.G., Lurie, S.A., Posypkin, M.A., Solyaev, Y.O.: Application of optimization methods for finding equilibrium states of two-dimensional crystals. Comput. Math. Math. Phys. **56**(12), 2001–2010 (2016)

85. Evtushenko, Y.G., Malkova, V.U., Stanevichyus, A.A.: Parallelization of the global extremum searching process. Autom. Remote Control **68**(5), 787–798 (2007)

86. Evtushenko, Y.G., Malkova, V.U., Stanevichyus, A.A.: Parallel global optimization of functions of several variables. Comput. Math. Math. Phys. **49**(2), 246–260 (2009)

87. Evtushenko, Y.G., Posypkin, M.A.: Coverings for global optimization of partial-integer nonlinear problems. Dokl. Math. **83**(2), 1–4 (2011)

88. Evtushenko, Y.G., Posypkin, M.A., Sigal, I.K.: A framework for parallel large-scale global optimization. Comp. Sci. Res. Dev. **23**(3–4), 211–215 (2009)

89. Famularo, D., Pugliese, P., Sergeyev, Y.D.: Test problems for Lipschitz univariate global optimization with multiextremal constraints. In: Dzemyda, G., Šaltenis, V., Žilinskas, A. (eds.) Stochastic and Global Optimization, pp. 93–109. Kluwer Academic Publishers (2002)

90. Famularo, D., Pugliese, P., Sergeyev, Y.D.: A global optimization technique for fixed-order control design. Int. J. Syst. Sci. **35**(7), 425–434 (2004)

91. Fasano, G., Roma, M.: A novel class of approximate inverse preconditioners for large positive definite linear systems in optimization. Comput. Optim. Appl. **65**(2), 399–429 (2016)

92. Fiacco, A.V., McCormick, G.P.: Nonlinear programming: Sequential unconstrained minimization techniques. Wiley, New York (1968). (Reprinted by SIAM Publications, 1990)

93. Finkel, D.E., Kelley, C.T.: Additive scaling and the DIRECT algorithm. J. Glob. Optim. **36**(4), 597–608 (2006)

94. Fletcher, R.: Practical Methods of Optimization. Wiley, New York (2000)

95. Floudas, C.A.: Deterministic Global Optimization: Theory, Algorithms, and Applications. Kluwer Academic Publishers, Dordrecht (2000)

96. Floudas, C.A., Akrotirianakis, I.G., Caratzoulas, S., Meyer, C.A., Kallrath, J.: Global optimization in the 21st century: advances and challenges. Comput. Chem. Eng. **29**(6), 1185–1202 (2005)

97. Floudas, C.A., Gounaris, C.E.: A review of recent advances in global optimization. J. Glob. Optim. **45**(1), 3–38 (2009)

98. Floudas, C.A., Pardalos, P.M.: A Collection of Test Problems for Constrained Global Optimization Algorithms, vol. 455. Springer, New York (1990)

99. Floudas, C.A., Pardalos, P.M. (eds.): Recent Advances in Global Optimization. Princeton University Press, Princeton (1992)

100. Floudas, C.A., Pardalos, P.M. (eds.): State of The Art in Global Optimization: Computational Methods and Applications. Kluwer Academic Publishers, Dordrecht (1996)

101. Floudas, C.A., Pardalos, P.M. (eds.): Optimization in Computational Chemistry and Molecular Biology: Local and Global Approaches. Kluwer Academic Publishers, Massachusetts (2000)

102. Floudas, C.A., Pardalos, P.M. (eds.): Encyclopedia of Optimization, (6 volumes) 2nd edn. Springer, Heidelberg (2009)

103. Floudas, C.A., Pardalos, P.M., Adjiman, C.S., Esposito, W., Gümüs, Z., Harding, S., Klepeis, J., Meyer, C., Schweiger, C.: Handbook of Test Problems in Local and Global Optimization. Kluwer Academic Publishers, Dordrecht (1999)

104. Fowler, K.R., Reese, J.P., Kees, C.E., Dennis Jr., J.E., Kelley, C.T., Miller, C.T., Audet, C., Booker, A.J., Couture, G., Darwin, R.W., Farthing, M.W., Finkel, D.E., Gablonsky, J.M., Gray, G., Kolda, T.G.: Comparison of derivative-free optimization methods for groundwater supply and hydraulic capture community problems. Adv. Water Res. **31**(5), 743–757 (2008)

105. Fuduli, A., Gaudioso, M., Giallombardo, G.: Minimizing nonconvex nonsmooth functions via cutting planes and proximity control. SIAM J. Optim. **14**(3), 743–756 (2004)

106. Gablonsky, J.M.: Modifications of the DIRECT algorithm. Ph.D. thesis, North Carolina State University, Raleigh, NC, USA (2001)

107. Gablonsky, J.M., Kelley, C.T.: A locally-biased form of the DIRECT algorithm. J. Glob. Optim. **21**(1), 27–37 (2001)

108. Gallagher, M., Yuan, B.: A general-purpose tunable landscape generator. IEEE Trans. Evol. Comput. **10**(5), 590–603 (2006)

109. Galperin, E.A.: The cubic algorithm. J. Math. Anal. Appl. **112**(2), 635–640 (1985)

110. Garnier, H., Wang, L. (eds.): Identification of Continuous-Time Models from Sampled Data. Springer, London (2008)

111. Gaviano, M., Kvasov, D.E., Lera, D., Sergeyev, Y.D.: Algorithm 829: software for generation of classes of test functions with known local and global minima for global optimization. ACM Trans. Math. Softw. **29**(4), 469–480 (2003)

112. Gaviano, M., Lera, D.: A complexity analysis of local search algorithms in global optimization. Optim. Methods Softw. **17**(1), 113–127 (2002)

113. Gaviano, M., Lera, D.: A global minimization algorithm for Lipschitz functions. Optim. Lett. **2**(1), 1–13 (2008)

114. Gaviano, M., Lera, D., Steri, A.M.: A local search method for continuous global optimization. J. Glob. Optim. **48**(1), 73–85 (2010)
115. Gergel, A.V., Grishagin, V.A., Strongin, R.G.: Development of the parallel adaptive multistep reduction method. Vestnik of Lobachevsky State University of Nizhni Novgorod **6**(1), 216–222 (2013). (In Russian)
116. Gergel, V.P.: A global search algorithm using derivatives. In: Neimark, Y.I. (ed.) Systems Dynamics and Optimization, pp. 161–178. NNGU Press, Nizhni Novgorod, Russia (1992). (In Russian)
117. Gergel, V.P.: A global optimization algorithm for multivariate function with Lipschitzian first derivatives. J. Glob. Optim. **10**(3), 257–281 (1997)
118. Gergel, V.P., Grishagin, V.A., Gergel, A.V.: Adaptive nested optimization scheme for multi-dimensional global search. J. Glob. Optim. **66**(1), 35–51 (2016)
119. Gergel, V.P., Grishagin, V.A., Israfilov, R.A.: Local tuning in nested scheme of global optimization. Proced. Comput. Sci. **51**, 865–874 (2015). (International Conference on Computational Science ICCS 2015—Computational Science at the Gates of Nature)
120. Gillard, J.W., Kvasov, D.E.: Lipschitz optimization methods for fitting a sum of damped sinusoids to a series of observations. Stat. Int. **10**, 59–70 (2017)
121. Gillard, J.W., Kvasov, D.E., Zhigljavsky, A.: Optimization problems in structured low rank optimization. In: Sergeyev, Y.D., et al. (ed.) Numerical Computations: Theory and Algorithms (NUMTA-2016). AIP Conference on Proceedings, vol. 1776, p. 060004. AIP Publishing, New York (2016)
122. Gillard, J.W., Zhigljavsky, A.: Analysis of structured low rank approximation as an optimisation problem. Informatica **22**(4), 489–505 (2011)
123. Gillard, J.W., Zhigljavsky, A.: Optimization challenges in the structured low rank approximation problem. J. Glob. Optim. **57**(3), 733–751 (2013)
124. Gillard, J.W., Zhigljavsky, A.: Application of structured low-rank approximation methods for imputing missing values in time series. Stat. Int. **8**(3), 321–330 (2015)
125. Gillard, J.W., Zhigljavsky, A.: Stochastic algorithms for solving structured low-rank matrix approximation problems. Commun. Nonlinear Sci. Numer. Simul. **21**(1–3), 70–88 (2015)
126. Golyandina, N., Nekrutkin, V., Zhigljavsky, A.: Analysis of Time Series Structure: SSA and Related Techniques. Chapman & Hall/CRC, Boca Raton (2001)
127. Gorodetsky, S.Y.: Multiextremal optimization based on domain triangulation, The Bulletin of Nizhni Novgorod "Lobachevsky" University. Math. Model. Optim. Control **2**(21), 249–268 (1999, in Russian)
128. Gorodetsky, S.Y.: Several approaches to generalization of the DIRECT method to problems with functional constraints, The Bulletin of Nizhni Novgorod "Lobachevsky" University. Math. Model. Optim. Control **6**(1), 189–215 (2013, in Russian)
129. Gorodetsky, S.Y., Grishagin, V.A.: Nonlinear Programming and Multiextremal Optimization, Models and Methods of Finite-Dimensional Optimization. NNGU Press, Nizhni Novgorod, Russia (2007, in Russian)
130. Gorodetsky, S.Y., Sorokin, A.S.: Constructing optimal controllers using nonlinear performance criteria on the example of one dynamic system, The Bulletin of Nizhni Novgorod "Lobachevsky" University. Math. Model. Optim. Control **2**(1), 165–176 (2012, in Russian)
131. Gourdin, E., Jaumard, B., Ellaia, R.: Global optimization of Hölder functions. J. Glob. Optim. **8**(4), 323–348 (1996)
132. Grishagin, V.A.: Operating characteristics of some global search algorithms. In: Problems of Stochastic Search, vol. 7, pp. 198–206. Zinatne, Riga (1978, in Russian)
133. Grishagin, V.A.: On convergence conditions for a class of global search algorithms. In: Proceedings of the 3rd All–Union Seminar on Numerical Methods of Nonlinear Programming, pp. 82–84. Kharkov (1979, in Russian)
134. Grishagin, V.A., Israfilov, R., Sergeyev, Y.D.: Comparative efficiency of dimensionality reduction schemes in global optimization. In: Sergeyev, Y.D., et al. (ed.) Numerical Computations: Theory and Algorithms (NUMTA-2016). AIP Conference Proceedings, vol. 1776, p. 060011. AIP Publishing, New York (2016)

135. Grishagin, V.A., Sergeyev, Y.D., Strongin, R.G.: Parallel characteristic algorithms for solving problems of global optimization. J. Glob. Optim. **10**(2), 185–206 (1997)
136. Grishagin, V.A., Strongin, R.G.: Optimization of multi-extremal functions subject to monotonically unimodal constraints. Eng. Cybern. **22**(5), 117–122 (1984)
137. Grossmann, I.E. (ed.): Global Optimization in Engineering Design. Kluwer Academic Publishers, Dordrecht (1996)
138. Gutmann, H.M.: A radial basis function method for global optimization. J. Glob. Optim. **19**(3), 201–227 (2001)
139. Hansen, E.R. (ed.): Global Optimization Using Interval Analysis, Pure and Applied Mathematics, vol. 165. M. Dekker, New York (1992)
140. Hansen, P., Jaumard, B.: Lipschitz optimization. In: Horst, R., Pardalos, P.M. (eds.) Handbook of Global Optimization, vol. 1, pp. 407–493. Kluwer Academic Publishers, Dordrecht (1995)
141. He, J., Watson, L.T., Ramakrishnan, N., Shaffer, C.A., Verstak, A., Jiang, J., Bae, K., Tranter, W.H.: Dynamic data structures for a direct search algorithm. Comput. Optim. Appl. **23**(1), 5–25 (2002)
142. Himmelblau, D.: Applied Nonlinear Programming. McGraw-Hill, New York (1972)
143. Hiriart-Urruty, J.B., Lemaréchal, C.: Convex Analysis and Minimization Algorithms (Parts I and II). Springer, Berlin (1996)
144. Holland, J.H.: Adaptation in Natural and Artificial Systems. The University of Michigan Press, USA (1975)
145. Horst, R.: Deterministic global optimization with partition sets whose feasibility is not known: application to concave minimization, reverse convex constraints, DC-programming, and Lipschitzian optimization. J. Optim. Theory Appl. **58**(1), 11–37 (1988)
146. Horst, R.: On generalized bisection of N-simpices. Math. Comp. **66**(218), 691–698 (1997)
147. Horst, R., Nast, M., Thoai, N.V.: New LP bound in multivariate Lipschitz optimization: theory and applications. J. Optim. Theory Appl. **86**(2), 369–388 (1995)
148. Horst, R., Pardalos, P.M. (eds.): Handbook of Global Optimization, vol. 1. Kluwer Academic Publishers, Dordrecht (1995)
149. Horst, R., Pardalos, P.M., Thoai, N.V.: Introduction to Global Optimization. Kluwer Academic Publishers, Dordrecht (1995). (The 2nd edition: Kluwer Academic Publishers, 2001)
150. Horst, R., Tuy, H.: Global Optimization-Deterministic Approaches. Springer, Berlin (1996)
151. Huyer, W., Neumaier, A.: Global optimization by multilevel coordinate search. J. Glob. Optim. **14**(4), 331–355 (1999)
152. Ivanov, V.V.: On optimal algorithms for the minimization of functions of certain classes. Cybernetics **4**, 81–94 (1972). In Russian
153. Jones, D.R.: The DIRECT global optimization algorithm. In: Floudas, C.A., Pardalos, P.M. (eds.) Encyclopedia of Optimization, vol. 1, pp. 431–440. Kluwer Academic Publishers, Dordrecht (2001)
154. Jones, D.R., Perttunen, C.D., Stuckman, B.E.: Lipschitzian optimization without the Lipschitz constant. J. Optim. Theory Appl. **79**(1), 157–181 (1993)
155. Jones, D.R., Schonlau, M., Welch, W.J.: Efficient global optimization of expensive black-box functions. J. Glob. Optim. **13**(4), 455–492 (1998)
156. Kearfott, R.B.: Rigorous Global Search: Continuous Problems. Kluwer Academic Publishers, Dordrecht (1996)
157. Kelley, C.T.: Iterative Methods for Optimization, Frontiers in Applied Mathematics, vol. 18. SIAM Publications, Philadelphia (1999)
158. Khamisov, O.V.: Finding roots of nonlinear equations using the method of concave support functions. Math. Notes **98**(3–4), 484–491 (2015)
159. Khamisov, O.V.: Global optimization of functions with a concave support minorant. Comp. Math. Math. Phys. **44**(9), 1473–1483 (2015)
160. Kirsch, N., Alibeji, N., Sharma, N.: Nonlinear model predictive control of functional electrical stimulation. Control Eng. Pract. **58**, 319–331 (2017)
161. Kiseleva, E.M., Stepanchuk, T.: On the efficiency of a global non-differentiable optimization algorithm based on the method of optimal set partitioning. J. Glob. Optim. **25**(2), 209–235 (2003)

162. Kolda, T.G., Lewis, R.M., Torczon, V.: Optimization by direct search: New perspectives on some classical and modern methods. SIAM Rev. **45**(3), 385–482 (2003)
163. Korotchenko, A.G.: An algorithm for seeking the maximum value of univariate functions. USSR Comput. Math. Math. Phys. **18**(3), 34–45 (1978)
164. Korotchenko, A.G.: An approximately optimal algorithm for finding an extremum for a certain class of functions. Comput. Math. Math. Phys. **36**(5), 577–584 (1996)
165. Korotchenko, A.G., Smoryakova, V.M.: On error estimation of extremum search algorithms in function classes defined by a piecewise linear majorant, The Bulletin of Nizhni Novgorod "Lobachevsky" University. Math. Model. Optim. Control **3**(1), 188–194 (2013, in Russian)
166. Kushner, H.J., Clark, D.S.: Stochastic Approximation Methods for Constrained and Unconstrained Systems, Applied mathematical sciences, vol. 26. Springer, New York (1978)
167. Kuzenkov, O.A., Grishagin, V.A.: Global optimization in Hilbert space. In: Simos, T.E., Tsitouras, C. (eds.) Numerical Analysis and Applied Mathematics (ICNAAM 2015). AIP Conference Proceedings, vol. 1738, p. 400007. AIP Publishing, New York (2016)
168. Kvasov, D.E.: Algoritmi diagonali di ottimizzazione globale Lipschitziana basati su una efficiente strategia di partizione. Bollettino U.M.I. **10-A (Serie VIII)**(2), 255–258 (2007)
169. Kvasov, D.E.: Diagonal numerical methods for solving Lipschitz global optimization problems. Bollettino U.M.I. **I (Serie IX)**(3), 857–871 (2008)
170. Kvasov, D.E.: Multidimensional Lipschitz global optimization based on efficient diagonal partitions, 4OR - Quart. J Oper. Res. **6**(4), 403–406 (2008)
171. Kvasov, D.E., Menniti, D., Pinnarelli, A., Sergeyev, Y.D., Sorrentino, N.: Tuning fuzzy power-system stabilizers in multi-machine systems by global optimization algorithms based on efficient domain partitions. Electr. Power Syst. Res. **78**(7), 1217–1229 (2008)
172. Kvasov, D.E., Mukhametzhanov, M.S.: One-dimensional global search: Nature-inspired versus Lipschitz methods. In: Simos, T.E., Tsitouras, C. (eds.) Numerical Analysis and Applied Mathematics (ICNAAM 2015). AIP Conference Proceedings, vol. 1738, p. 400012. AIP Publishing, New York (2016)
173. Kvasov, D.E., Mukhametzhanov, M.S.: Metaheuristic vs. deterministic global optimization algorithms: The univariate case, Appl. Math. Comput. (2017). In Press: doi:10.1016/j.amc.2017.05.014
174. Kvasov, D.E., Mukhametzhanov, M.S., Sergeyev, Y.D.: A numerical comparison of some deterministic and nature-inspired algorithms for black-box global optimization. In: Topping, B.H.V., Iványi, P. (eds.) Proceedings of the Twelfth International Conference on Computational Structures Technology, vol. 106, p. 169. Civil-Comp Press, Stirlingshire, United Kingdom (2014). doi:10.4203/ccp.106.169
175. Kvasov, D.E., Pizzuti, C., Sergeyev, Y.D.: Local tuning and partition strategies for diagonal GO methods. Numer. Math. **94**(1), 93–106 (2003)
176. Kvasov, D.E., Sergeyev, Y.D.: Multidimensional global optimization algorithm based on adaptive diagonal curves. Comput. Math. Math. Phys. **43**(1), 40–56 (2003)
177. Kvasov, D.E., Sergeyev, Y.D.: A univariate global search working with a set of Lipschitz constants for the first derivative. Optim. Lett. **3**(2), 303–318 (2009)
178. Kvasov, D.E., Sergeyev, Y.D.: Deterministic gobal optimization methods for solving engineering problems. In: Topping, B.H.V. (ed.) Proceedings of the Eleventh International Conference on Computational Structures Technology, vol. 92, p. 62. Civil-Comp Press, Stirlingshire, United Kingdom (2012). doi:10.4203/ccp.99.62
179. Kvasov, D.E., Sergeyev, Y.D.: Lipschitz gradients for global optimization in a one-point-based partitioning scheme. J. Comput. Appl. Math. **236**(16), 4042–4054 (2012)
180. Kvasov, D.E., Sergeyev, Y.D.: Univariate geometric Lipschitz global optimization algorithms. Numer. Algebr. Contr. Optim. **2**(1), 69–90 (2012)
181. Kvasov, D.E., Sergeyev, Y.D.: Lipschitz global optimization methods in control problems. Autom. Remote Control **74**(9), 1435–1448 (2013)
182. Kvasov, D.E., Sergeyev, Y.D.: Deterministic approaches for solving practical black-box global optimization problems. Adv. Eng. Softw. **80**, 58–66 (2015)

183. Kvasov, I.E., Leviant, V.B., Petrov, I.B.: Solution to Direct Problems of Seismic Exploration in Fractured Subsurface Using Grid-Characteristic Modeling Method. Eage, Moscow (2016). In Russian

184. Lavor, C., Maculan, N.: A function to test methods applied to global minimization of potential energy of molecules. Numer. Algorithms 35(2–4), 287–300 (2004)

185. Lera, D., Sergeyev, Y.D.: Global minimization algorithms for Hölder functions. BIT 42(1), 119–133 (2002)

186. Lera, D., Sergeyev, Y.D.: An information global minimization algorithm using the local improvement technique. J. Glob. Optim. 48(1), 99–112 (2010)

187. Lera, D., Sergeyev, Y.D.: Lipschitz and Hölder global optimization using space-filling curves. Appl. Numer. Math. 60(1–2), 115–129 (2010)

188. Lera, D., Sergeyev, Y.D.: Acceleration of univariate global optimization algorithms working with Lipschitz functions and Lipschitz first derivatives. SIAM J. Optim. 23(1), 508–529 (2013)

189. Lera, D., Sergeyev, Y.D.: Deterministic global optimization using space-filling curves and multiple estimates of Lipschitz and Hölder constants. Commun. Nonlinear Sci. Numer. Simul. 23(1–3), 328–342 (2015)

190. Li, Y., Liu, K.J.R., Razavilar, J.: A parameter estimation scheme for damped sinusoidal signals based on low-rank Hankel approximation. IEEE Trans. Signal Process. 45(2), 481–486 (1997)

191. Liberti, L., Maculan, N. (eds.): Global Optimization: From Theory to Implementation, Nonconvex Optimization and Its Applications, vol. 84. Springer, Berlin (2006)

192. Liu, Q.: Linear scaling and the DIRECT algorithm. J. Glob. Optim. 56, 1233–1245 (2013)

193. Liu, Q., Cheng, W.: A modified DIRECT algorithm with bilevel partition. J. Glob. Optim. 60(3), 483–499 (2014)

194. Liuzzi, G., Lucidi, S., Piccialli, V.: A DIRECT-based approach exploiting local minimizations for the solution of large-scale global optimization problems. Comput. Optim. Appl. 45(2), 353–375 (2010)

195. Liuzzi, G., Lucidi, S., Piccialli, V.: A partition-based global optimization algorithm. J. Glob. Optim. 48, 113–128 (2010)

196. Liuzzi, G., Lucidi, S., Piccialli, V., Sotgiu, A.: A magnetic resonance device designed via global optimization techniques. Math. Progr. 101(2), 339–364 (2004)

197. Liuzzi, G., Lucidi, S., Rinaldi, F.: A derivative-free approach to constrained multiobjective nonsmooth optimization. SIAM J. Optim. 26(4), 2744–2774 (2016)

198. Locatelli, M.: On the multilevel structure of global optimization problems. Comput. Optim. Appl. 30(1), 5–22 (2005)

199. Locatelli, M., Schoen, F.: Global Optimization: Theory, Algorithms, and Applications, MOS-SIAM Series on Optimization, vol. 282. SIAM, USA (2013)

200. MacLagan, D., Sturge, T., Baritompa, W.: Equivalent methods for global optimization. In: Floudas, C.A., Pardalos, P.M. (eds.) State of the Art in Global Optimization, pp. 201–211. Kluwer Academic Publishers, Dordrecht (1996)

201. Mangasarian, O.L.: Nonlinear Programming. McGraw–Hill, New York (1969). (Reprinted by SIAM Publications, 1994)

202. Maranas, C.D., Floudas, C.A.: Global minimum potential energy conformations of small molecules. J. Glob. Optim. 4(2), 135–170 (1994)

203. Mayne, D.Q., Polak, E.: Outer approximation algorithm for nondifferentiable optimization problems. J. Optim. Theory Appl. 42(1), 19–30 (1984)

204. Meewella, C.C., Mayne, D.Q.: An algorithm for global optimization of Lipschitz continuous functions. J. Optim. Theory Appl. 57(2), 307–322 (1988)

205. Meewella, C.C., Mayne, D.Q.: Efficient domain partitioning algorithms for global optimization of rational and Lipschitz continuous functions. J. Optim. Theory Appl. 61(2), 247–270 (1989)

206. Michalewicz, Z.: Genetic Algorithms + Data Structures = Evolution Programs, 3rd edn. Springer, Berlin (1996)

207. Mladineo, R.H.: An algorithm for finding the global maximum of a multimodal multivariate function. Math. Progr. 34(2), 188–200 (1986)

208. Mladineo, R.H.: Convergence rates of a global optimization algorithm. Math. Progr. **54**(1–3), 223–232 (1992)
209. Mockus, J.: Multiextremal Problems in Engineering Design. Nauka, Moscow (1967). In Russian
210. Mockus, J.: Bayesian Approach to Global Optimization. Kluwer Academic Publishers, Dordrecht (1989)
211. Mockus, J.: A Set of Examples of Global and Discrete Optimization: Applications of Bayesian Heuristic Approach. Kluwer Academic Publishers, Dordrecht (2000)
212. Mockus, J., Eddy, W., Mockus, A., Mockus, L., Reklaitis, G.: Bayesian Heuristic Approach to Discrete and Global Optimization. Kluwer Academic Publishers, Dordrecht (1996)
213. Mockus, J., Paulavičius, R., Rusakevičius, D., Šešok, D., Žilinskas, J.: Application of Reduced-set Pareto-Lipschitzian Optimization to truss optimization. J. Glob. Optim. **67**(1–2), 425–450 (2017)
214. Modorskii, V.Y., Gaynutdinova, D.F., Gergel, V.P., Barkalov, K.A.: Optimization in design of scientific products for purposes of cavitation problems. In: Simos, T.E., Tsitouras, C. (eds.) Numerical Analysis and Applied Mathematics (ICNAAM 2015). AIP Conference Proceedings, vol. 1738, p. 400013. AIP Publishing, New York (2016)
215. Moles, C.G., Mendes, P., Banga, J.R.: Parameter estimation in biochemical pathways: a comparison of global optimization methods. Gen. Res. **13**(11), 2467–2474 (2003)
216. Molinaro, A., Pizzuti, C., Sergeyev, Y.D.: Acceleration tools for diagonal information global optimization algorithms. Comput. Optim. Appl. **18**(1), 5–26 (2001)
217. Molinaro, A., Sergeyev, Y.D.: An efficient algorithm for the zero-crossing detection in digitized measurement signal. Measurement **30**(3), 187–196 (2001)
218. Molinaro, A., Sergeyev, Y.D.: Finding the minimal root of an equation with the multiextremal and nondifferentiable left-hand part. Numer. Algor. **28**(1–4), 255–272 (2001)
219. Nefedov, V.N.: Some problems of solving Lipschitzian global optimization problems using the branch and bound method. Comput. Math. Math. Phys. **32**(4), 433–445 (1992)
220. Neimark, Y.I., Strongin, R.G.: The information approach to the problem of search of extrema of functions. Eng. Cybern. **1**, 17–26 (1966)
221. Nesterov, Y.: Introductory Lectures on Convex Optimization: A Basic Course. Kluwer Academic Publishers, Dordrecht (2004)
222. Neumaier, A., Shcherbina, O., Huyer, W., Vinkó, T.: A comparison of complete global optimization solvers. Math. Progr. **103**(2), 335–356 (2005)
223. Nocedal, J., Wright, S.J.: Numerical Optimization. Springer, Dordrecht (1999)
224. Norkin, V.I.: On Piyavskij's method for solving the general global optimization problem. Comput. Math. Math. Phys. **32**(7), 873–886 (1992)
225. Pardalos, P.M. (ed.): Approximation and Complexity in Numerical Optimization: Continuous and Discrete Problems. Kluwer Academic Publishers, Dordrecht (2000)
226. Pardalos, P.M., Philips, A.T., Rosen, J.B.: Topics in Parallel Computing in Mathematical Programming. Science Press, New York (1992)
227. Pardalos, P.M., Resende, M.G.C. (eds.): Handbook of Applied Optimization. Oxford University Press, New York (2002)
228. Pardalos, P.M., Romeijn, H.E. (eds.): Handbook of Global Optimization, vol. 2. Kluwer Academic Publishers, Dordrecht (2002)
229. Pardalos, P.M., Romeijn, H.E. (eds.): Handbook of Optimization in Medicine. Springer, New York (2009)
230. Pardalos, P.M., Romeijn, H.E., Tuy, H.: Recent developments and trends in global optimization. J. Comput. Appl. Math. **124**(1–2), 209–228 (2000)
231. Pardalos, P.M., Rosen, J.B.: Constrained Global Optimization: Algorithms and Applications. Springer Lecture Notes In Computer Science, vol. 268. Springer, New York (1987)
232. Paulavičius, R., Chiter, L., Žilinskas, J.: Global optimization based on bisection of rectangles, function values at diagonals, and a set of Lipschitz constants. J. Glob. Optim. (2017). doi:10.1007/s10898-016-0485-6

233. Paulavičius, R., Sergeyev, Y.D., Kvasov, D.E., Žilinskas, J.: Globally-biased DISIMPL algorithm for expensive global optimization. J. Glob. Optim. **59**(2–3), 545–567 (2014)
234. Paulavičius, R., Žilinskas, J., Grothey, A.: Investigation of selection strategies in branch and bound algorithm with simplicial partitions and combination of Lipschitz bounds. Optim. Lett. **4**(2), 173–183 (2010)
235. Paulavičius, R., Žilinskas, J.: Simplicial Global Optimization. SpringerBriefs in Optimization. Springer, New York (2014)
236. Paulavičius, R., Žilinskas, J.: Simplicial Lipschitz optimization without the Lipschitz constant. J. Glob. Optim. **59**(1), 23–40 (2014)
237. Paulavičius, R., Žilinskas, J.: Advantages of simplicial partitioning for Lipschitz optimization problems with linear constraints. Optim. Lett. **10**(2), 237–246 (2016)
238. Peitgen, H.O., Jürgens, H., Saupe, D.: Chaos and Fractals: New Frontiers of Science. Springer, New York (1992)
239. Pintér, J.D.: A unified approach to globally convergent one-dimensional optimization algorithms. Technical Report 83–5, IAMI–CNR, Institute of Applied Mathematics and Informatics CNR, Milan, Italy (1983)
240. Pintér, J.D.: Extended univariate algorithms for N-dimensional global optimization. Computing **36**(1–2), 91–103 (1986)
241. Pintér, J.D.: Convergence qualification of adaptive partition algorithms in global optimization. Math. Progr. **56**(1–3), 343–360 (1992)
242. Pintér, J.D.: Global Optimization in Action (Continuous and Lipschitz Optimization: Algorithms, Implementations and Applications). Kluwer Academic Publishers, Dordrecht (1996)
243. Pintér, J.D.: Global optimization: software, test problems, and applications. In: Pardalos, P.M., Romeijn, H.E. (eds.) Handbook of Global Optimization, vol. 2, pp. 515–569. Kluwer Academic Publishers, Dordrecht (2002)
244. Pintér, J.D. (ed.): Global Optimization: Scientific and Engineering Case Studies, Nonconvex Optimization and Its Applications, vol. 85. Springer, Berlin (2006)
245. Piyavskij, S.A.: An algorithm for finding the absolute minimum of a function. In: Optimum Decison Theory, vol. 2, pp. 13–24. Inst. Cybern. Acad. Science Ukrainian SSR, Kiev (1967, in Russian)
246. Piyavskij, S.A.: An algorithm for finding the absolute extremum of a function. USSR Comput. Math. Math. Phys. **12**(4), 57–67 (1972). (In Russian: Zh. Vychisl. Mat. Mat. Fiz., 12(4) (1972), pp. 888–896)
247. Pollock, D.S.G.: A Handbook of Time Series Analysis, Signal Processing, and Dynamics. Academic Press Inc., London (1999)
248. Posypkin, M.A.: Method for solving constrained multicriteria optimization problems with guaranteed accuracy. Dokl. Math. **88**(2), 559–561 (2013)
249. Posypkin, M.A., Usov, A.L.: Visualizing of large trees, resulting in solving optimization problems by branch and bound algorithm. Int. J. Open Inf. Tech. **4**(8), 43–49 (2016)
250. Preparata, F.P., Shamos, M.I.: Computational Geometry: An Introduction. Monographs in Computer Science. Springer, New York (1993)
251. Price, K., Storn, R., Lampinen, J.: Differential Evolution: A Practical Approach to Global Optimization. Springer, New York (2005)
252. Rastrigin, L.A.: Adaptation of Complex Systems: Methods and Applications. Zinatne, Riga (1981, in Russian)
253. Ratschek, H., Rockne, J.: New Computer Methods for Global Optimization. Mathematics and Its Applications. Ellis Horwood Ltd, Chichester, England (1988)
254. Ratz, D., Csendes, T.: On the selection of subdivision directions in interval branch-and-bound methods for global optimization. J. Glob. Optim. **7**(2), 183–207 (1995)
255. Rebennack, S., Pardalos, P.M., Pereira, M.V.F., Iliadis, N.A. (eds.): Handbook of Power Systems I. Spinger, New York (2010)
256. Regis, R.G., Shoemaker, C.A.: Constrained global optimization of expensive black box functions using radial basis functions. J. Glob. Optim. **31**(1), 153–171 (2005)

257. Resende, M.G.C., Pardalos, P.M. (eds.): Handbook of Optimization in Telecommunications. Spinger, New York (2006)
258. Rinnooy Kan, A.H.G., Timmer, G.T.: Global optimization. In: Nemhauser, G.L., Rinnooy Kan, A.H.G., Todd, M.J. (eds.) Handbook of Operations Research Optimization, vol. 1, pp. 631–662. North–Holland, Amsterdam (1989)
259. Rios, L.M., Sahinidis, N.V.: Derivative-free optimization: a review of algorithms and comparison of software implementations. J. Glob. Optim. **56**, 1247–1293 (2013)
260. Rockafellar, R.T.: Convex Analysis. Princeton University Press, Princeton (1970)
261. Ruggiero, V., Serafini, T., Zanella, R., Zanni, L.: Iterative regularization algorithms for constrained image deblurring on graphics processors. J. Glob. Optim. **48**(1), 145–157 (2010)
262. Saez-Landete, J., Alonso, J., Bernabeu, E.: Design of zero reference codes by means of a global optimization method. Optics Express **13**(1), 195–201 (2005)
263. Sagan, H.: Space-Filling Curves. Springer, New York (1994)
264. Schittkowski, K.: More Test Examples for Nonlinear Programming Codes. Lecture Notes in Economics and Mathematical Systems, vol. 282. Springer, Berlin (1987)
265. Schneider, J.J., Kirkpatrick, S.: Stochastic Optimization. Springer, Berlin (2006)
266. Schoen, F.: On a sequential search strategy in global optimization problems. Calcolo **19**, 321–334 (1982)
267. Schoen, F.: Stochastic techniques for global optimization: a survey of recent advances. J. Glob. Optim. **1**(1), 207–228 (1991)
268. Schoen, F.: A wide class of test functions for global optimization. J. Glob. Optim. **3**(2), 133–137 (1993)
269. Schwefel, H.P.: Evolution and Optimum Seeking. Wiley, New York (1995)
270. Sergeyev, Y.D.: A global optimization algorithm using derivatives and local tuning. Technical Report 1, ISI–CNR, Institute of Systems and Informatics, Rende(CS), Italy (1994)
271. Sergeyev, Y.D.: Global optimization algorithms using smooth auxiliary functions. Technical Report 5, ISI–CNR, Institute of Systems and Informatics, Rende(CS), Italy (1994)
272. Sergeyev, Y.D.: An information global optimization algorithm with local tuning. SIAM J. Optim. **5**(4), 858–870 (1995)
273. Sergeyev, Y.D.: A one-dimensional deterministic global minimization algorithm. Comput. Math. Math. Phys. **35**(5), 705–717 (1995)
274. Sergeyev, Y.D.: A two-points-three-intervals partition of the N-dimensional hyperinterval. Technical Report 10, ISI–CNR, Institute of Systems and Informatics, Rende(CS), Italy (1995)
275. Sergeyev, Y.D.: A method using local tuning for minimizing functions with Lipschitz derivatives. In: Bomze, I.M., Csendes, T., Horst, R., Pardalos, P.M. (eds.) Developments in Global Optimization, pp. 199–216. Kluwer Academic Publishers (1997)
276. Sergeyev, Y.D.: Global one-dimensional optimization using smooth auxiliary functions. Math. Progr. **81**(1), 127–146 (1998)
277. Sergeyev, Y.D.: On convergence of "Divide the Best" global optimization algorithms. Optimization **44**(3), 303–325 (1998)
278. Sergeyev, Y.D.: Multidimensional global optimization using the first derivatives. Comput. Math. Math. Phys. **39**(5), 711–720 (1999)
279. Sergeyev, Y.D.: An efficient strategy for adaptive partition of N-dimensional intervals in the framework of diagonal algorithms. J. Optim. Theory Appl. **107**(1), 145–168 (2000)
280. Sergeyev, Y.D.: Efficient partition of N-dimensional intervals in the framework of one-point-based algorithms. J. Optim. Theory Appl. **124**(2), 503–510 (2005)
281. Sergeyev, Y.D.: Univariate global optimization with multiextremal non-differentiable constraints without penalty functions. Comput. Optim. Appl. **34**(2), 229–248 (2006)
282. Sergeyev, Y.D., Daponte, P., Grimaldi, D., Molinaro, A.: Two methods for solving optimization problems arising in electronic measurements and electrical engineering. SIAM J. Optim. **10**(1), 1–21 (1999)
283. Sergeyev, Y.D., Famularo, D., Pugliese, P.: Index Branch-and-Bound Algorithm for Lipschitz univariate global optimization with multiextremal constraints. J. Glob. Optim. **21**(3), 317–341 (2001)

284. Sergeyev, Y.D., Grishagin, V.A.: A parallel method for finding the global minimum of univariate functions. J. Optim. Theory Appl. **80**(3), 513–536 (1994)
285. Sergeyev, Y.D., Grishagin, V.A.: Sequential and parallel algorithms for global optimization. Optim. Meth. Softw. **3**(1–3), 111–124 (1994)
286. Sergeyev, Y.D., Grishagin, V.A.: Parallel asynchronous global search and the nested optimization scheme. J. Comput. Anal. Appl. **3**(2), 123–145 (2001)
287. Sergeyev, Y.D., Khalaf, F.M.H., Kvasov, D.E.: Univariate algorithms for solving global optimization problems with multiextremal non-differentiable constraints. In: Törn, A., Žilinskas, J. (eds.) Models and Algorithms for Global Optimization, pp. 123–140. Springer, New York (2007)
288. Sergeyev, Y.D., Kvasov, D.E.: Adaptive diagonal curves and their implementation, The Bulletin of Nizhni Novgorod "Lobachevsky" University. Math. Model. Optim. Control **2**(24), 300–317 (2001, in Russian)
289. Sergeyev, Y.D., Kvasov, D.E.: Global search based on efficient diagonal partitions and a set of Lipschitz constants. SIAM J. Optim. **16**(3), 910–937 (2006)
290. Sergeyev, Y.D., Kvasov, D.E.: Diagonal Global Optimization Methods. FizMatLit, Moscow (2008, in Russian)
291. Sergeyev, Y.D., Kvasov, D.E.: Lipschitz global optimization. In: Cochran, J.J., Cox, L.A., Keskinocak, P., Kharoufeh, J.P., Smith, J.C. (eds.) Wiley Encyclopedia of Operations Research and Management Science (in 8 volumes), vol. 4, pp. 2812–2828. Wiley, New York (2011)
292. Sergeyev, Y.D., Kvasov, D.E.: A deterministic global optimization using smooth diagonal auxiliary functions. Commun. Nonlinear Sci. Numer. Simul. **21**(1–3), 99–111 (2015)
293. Sergeyev, Y.D., Kvasov, D.E.: On deterministic diagonal methods for solving global optimization problems with Lipschitz gradients. In: Migdalas, A., Karakitsiou, A. (eds.) Optimization, Control, and Applications in the Information Age, Springer Proceedings in Mathematics & Statistics, vol. 130, pp. 315–334. Springer, Cham (2015)
294. Sergeyev, Y.D., Kvasov, D.E., Khalaf, F.M.H.: A one-dimensional local tuning algorithm for solving GO problems with partially defined constraints. Optim. Lett. **1**(1), 85–99 (2007)
295. Sergeyev, Y.D., Kvasov, D.E., Mukhametzhanov, M.S.: Comments upon the usage of derivatives in Lipschitz global optimization. In: Simos, T.E., Tsitouras, C. (eds.) Numerical Analysis and Applied Mathematics (ICNAAM 2015). AIP Conference Proceedings, vol. 1738, p. 400004. AIP Publishing, New York (2016)
296. Sergeyev, Y.D., Kvasov, D.E., Mukhametzhanov, M.S.: On the least-squares fitting of data by sinusoids. In: Pardalos, P.M., Zhigljavsky, A., Žilinskas, J. (eds.) Advances in Stochastic and Deterministic Global Optimization, Springer Optimization and Its Applications, vol. 107, pp. 209–226. Springer, Switzerland (2016)
297. Sergeyev, Y.D., Kvasov, D.E., Mukhametzhanov, M.S.: Emmental-type GKLS-based multiextremal smooth test problems with non-linear constraints. In: Learning and Intelligent Optimization Conference (LION 2017). LNCS. Springer (2017). (To appear)
298. Sergeyev, Y.D., Kvasov, D.E., Mukhametzhanov, M.S.: Operational zones for comparing metaheuristic and deterministic one-dimensional global optimization algorithms. Math. Comp. Simul. (2017). doi:10.1016/j.matcom.2016.05.006
299. Sergeyev, Y.D., Kvasov, D.E., Mukhametzhanov, M.S., De Franco, A.: Acceleration techniques in the univariate Lipschitz global optimization. In: Sergeyev, Y.D., et al. (ed.) Numerical Computations: Theory and Algorithms (NUMTA-2016). AIP Conference Proceedings, vol. 1776, p. 090051. AIP Publishing, New York (2016)
300. Sergeyev, Y.D., Markin, D.L.: An algorithm for solving global optimization problems with nonlinear constraints. J. Glob. Optim. **7**(4), 407–419 (1995)
301. Sergeyev, Y.D., Mukhametzhanov, M.S., Kvasov, D.E., Lera, D.: Derivative-free local tuning and local improvement techniques embedded in the univariate global optimization. J. Optim. Theory Appl. **171**(1), 186–208 (2016)
302. Sergeyev, Y.D., Pugliese, P., Famularo, D.: Index information algorithm with local tuning for solving multidimensional global optimization problems with multiextremal constraints. Math. Progr. **96**(3), 489–512 (2003)

303. Sergeyev, Y.D., Strongin, R.G., Lera, D.: Introduction to Global Optimization Exploiting Space-Filling Curves. SpringerBriefs in Optimization. Springer, New York (2013)

304. Shen, Z., Zhu, Y.: An interval version of Shubert's iterative method for the localization of the global maximum. Computing **38**(3), 275–280 (1987)

305. Sherali, H.D., Ganesan, V.: A pseudo-global optimization approach with application to the design of containerships. J. Glob. Optim. **26**(4), 335–360 (2003)

306. Shubert, B.O.: A sequential method seeking the global maximum of a function. SIAM J. Numer. Anal. **9**(3), 379–388 (1972)

307. Spedicato, E.: Algorithms for Continuous Optimization: The State of the Art, NATO Science Series C, vol. 434. Kluwer Academic Publishers, Dordrecht (1994)

308. Stephens, C.P., Baritompa, W.: Global optimization requires global information. J. Optim. Theory Appl. **96**(3), 575–588 (1998)

309. Strekalovsky, A.S.: Global optimality conditions for nonconvex optimization. J. Glob. Optim. **12**(4), 415–434 (1998)

310. Strekalovsky, A.S.: Elements of Nonconvex Optimization. Nauka, Novosibirsk (2003). In Russian

311. Strekalovsky, A.S., Orlov, A.V.: Bimatrix Games and Bilinear Programming. Fizmatlit, Moscow (2007). In Russian

312. Strigul, O.I.: Search for a global extremum in a certain subclass of functions with the Lipschitz condition. Cybernetics **6**, 72–76 (1985). In Russian

313. Strongin, R.G.: Multiextremal minimization for measurements with interference. Eng. Cybern. **16**, 105–115 (1969)

314. Strongin, R.G.: On the convergence of an algorithm for finding a global extremum. Eng. Cybern. **11**, 549–555 (1973)

315. Strongin, R.G.: Numerical Methods in Multiextremal Problems (Information-Statistical Algorithms). Nauka, Moscow (1978). In Russian

316. Strongin, R.G.: Numerical methods for multiextremal nonlinear programming problems with nonconvex constraints. In: Demyanov, V.F., Pallaschke, D. (eds.) Nondifferentiable Optimization: Motivations and Applications. Proceedings, 1984. Lecture Notes in Economics and Mathematical Systems, vol. 255, pp. 278–282. Springer, IIASA, Laxenburg (1985)

317. Strongin, R.G.: Search for Global Optimum, Mathematics and Cybernetics, vol. 2. Znanie, Moscow (1990). In Russian

318. Strongin, R.G.: Algorithms for multi-extremal mathematical programming problems employing the set of joint space-filling curves. J. Glob. Optim. **2**(4), 357–378 (1992)

319. Strongin, R.G.: Global optimization using space filling. In: Floudas, C.A., Pardalos, P.M. (eds.) Encyclopedia of Optimization, vol. 2, pp. 345–350. Kluwer, Dordrecht (2001)

320. Strongin, R.G., Gergel, V.P., Grishagin, V.A., Barkalov, K.A.: Parallel Computing for Global Optimization Problems. Moscow University Press, Moscow (2013). In Russian

321. Strongin, R.G., Malkin, D.L.: Minimization of multiextremal functions with nonconvex constraints. Cybernetics **22**, 486–493 (1986)

322. Strongin, R.G., Sergeyev, Y.D.: Global multidimensional optimization on parallel computer. Parallel Comput. **18**(11), 1259–1273 (1992)

323. Strongin, R.G., Sergeyev, Y.D.: Global Optimization with Non-Convex Constraints: Sequential and Parallel Algorithms. Kluwer Academic Publishers, Dordrecht (2000). 3rd ed. by Springer (2014)

324. Strongin, R.G., Sergeyev, Y.D.: Global optimization: Fractal approach and non-redundant parallelism. J. Glob. Optim. **27**(1), 25–50 (2003)

325. Sukharev, A.G.: Best sequential search strategies for finding an extremum. USSR Comput. Math. Math. Phys. **12**(1), 39–59 (1972)

326. Sukharev, A.G.: Minimax Algorithms in Problems of Numerical Analysis. Nauka, Moscow (1989). In Russian

327. Tawarmalani, M., Sahinidis, N.V.: Convexification and Global Optimization in Continuous and Mixed-Integer Nonlinear Programming: Theory, Algorithms, Software, and Applications. Kluwer Academic Publishers, Dordrecht (2002)

328. Timonov, L.N.: An algorithm for search of a global extremum. Eng. Cybern. **15**, 38–44 (1977)
329. Törn, A., Žilinskas, A.: Global Optimization. Lecture Notes in Computer Science, vol. 350. Springer, Berlin (1989)
330. Tuy, H.: D. C. Optimization and robust global optimization. J. Glob. Optim. **47**(3), 485–501 (2010)
331. van Dam, E.R., Husslage, B., den Hertog, D.: One-dimensional nested maximin designs. J. Glob. Optim. **46**(2), 287–306 (2010)
332. Van Laarhoven, P.J.M., Aarts, E.H.L.: Simulated Annealing: Theory and Applications. Kluwer Academic Publishers, Dordrecht (1987)
333. Vanderbei, R.J.: Extension of Piyavskii's algorithm to continuous global optimization. J. Glob. Optim. **14**(2), 205–216 (1999)
334. Vasile, M., Locatelli, M.: A hybrid multiagent approach for global trajectory optimization. J. Glob. Optim. **44**(4), 461–479 (2009)
335. Vasile, M., Summerer, L., De Pascale, P.: Design of Earth-Mars transfer trajectories using evolutionary-branching technique. Acta Astronaut. **56**(8), 705–720 (2005)
336. Vasiliev, F.P.: Numerical Methods for Solving Extremum Problems. Nauka, Moscow (1988). In Russian
337. Villemonteix, J., Vazquez, E., Walter, E.: An informational approach to the global optimization of expensive-to-evaluate functions. J. Glob. Optim. **44**(4), 509–534 (2009)
338. Watson, L.T., Baker, C.: A fully-distributed parallel global search algorithm. Eng. Comput. **18**(1–2), 155–169 (2001)
339. Wild, S.M., Shoemaker, C.: Global convergence of radial basis function trust-region algorithms for derivative-free optimization. SIAM Rev. **55**(2), 349–371 (2013)
340. Wolpert, D.H., Macready, W.G.: No free lunch theorems for optimization. IEEE Trans. Evolut. Comput. **1**(1), 67–82 (1997)
341. Wood, G.R.: The bisection method in higher dimensions. Math. Progr. **55**(1–3), 319–337 (1992)
342. Wood, G.R., Zhang, B.: Estimation of the Lipschitz constant of a function. J. Glob. Optim. **8**(1), 91–103 (1996)
343. Yang, X.-S.: Engineering Optimization: An Introduction with Metaheuristic Applications. Wiley, USA (2010)
344. Baoping, Zhang: Wood, G.R., Baritompa, W.: Multidimensional bisection: the performance and the context. J. Glob. Optim. **3**(3), 337–358 (1993)
345. Zhigljavsky, A., Hamilton, E.: Stopping rules in k-adaptive global random search algorithms. J. Glob. Optim. **48**(1), 87–97 (2010)
346. Zhigljavsky, A.A.: Theory of Global Random Search. Kluwer Academic Publishers, Dordrecht (1991)
347. Zhigljavsky, A.A., Žilinskas, A.: Methods for Searching the Global Extremum. Nauka, Moscow (1991). In Russian
348. Zhigljavsky, A.A., Žilinskas, A.: Stochastic Global Optimization. Springer, New York (2008)
349. Žilinskas, A.: Axiomatic approach to statistical models and their use in multimodal optimization theory. Math. Progr. **22**(1), 104–116 (1982)
350. Žilinskas, A.: Glob. Optim. Algorithms, and Applications. Mokslas, Vilnius, Axiomatics of Statistical Models (1986, in Russian)
351. Žilinskas, A.: On similarities between two models of global optimization: statistical models and radial basis functions. J. Glob. Optim. **48**(1), 173–182 (2010)
352. Žilinskas, A.: On strong homogeneity of two global optimization algorithms based on statistical models of multimodal objective functions. Appl. Math. Comput. **218**(16), 8131–8136 (2012)
353. Žilinskas, A.: Visualization of a statistical approximation of the Pareto front. Appl. Math. Comput. **271**, 694–700 (2015)
354. Žilinskas, A., Gimbutiene, G.: On one-step worst-case optimal trisection in univariate bi-objective Lipschitz optimization. Commun. Nonlinear Sci. Numer. Simul. **35**(1), 123–136 (2016)

355. Žilinskas, A., Zhigljavsky, A.: Branch and probability bound methods in multi-objective optimization. Optim. Lett. **10**(2), 341–353 (2016)
356. Žilinskas, A., Zhigljavsky, A.: Stochastic global optimization: a review on the occasion of 25 years of Informatica. Informatica **27**(2), 229–256 (2016)
357. Žilinskas, A., Žilinskas, J.: Global optimization based on a statistical model and simplicial partitioning. Comput. Math. Appl. **44**(7), 957–967 (2002)

Printed in the United States
By Bookmasters